HOW TO
DVJ

A DIGITAL DJ MASTERCLASS

CHARLES KRIEL

www.howtodvj.com

ELSEVIER

AMSTERDAM · BOSTON · HEIDELBERG · LONDON · NEW YORK · OXFORD
PARIS · SAN DIEGO · SAN FRANCISCO · SINGAPORE · SYDNEY · TOKYO
Focal Press is an imprint of Elsevier

Focal Press is an imprint of Elsevier
Linacre House, Jordan Hill, Oxford OX2 8DP, UK
30 Corporate Drive, Suite 400, Burlington, MA 01803, USA

First edition 2007
Copyright © 2007, MIM. Published by Elsevier Ltd. All rights reserved

Notice
No responsibility is assumed by the publisher for any injury and/or damage to persons or property as a matter of products liability, negligence or otherwise, or from any use or operation of any methods, products, instructions or ideas contained in the material herein. Because of rapid advances in the medical sciences, in particular, independent verification of diagnoses and drug dosages should be made

British Library Cataloguing in Publication Data
A catalogue record for this book is available from the British Library

Library of Congress Cataloging-in-Publication Data
A catalog record for this book is available from the Library of Congress

ISBN–13: 978-0-240-52074-2

For information on all Focal Press publications
visit our website at www.focalpress.com

Printed and bound in China
07 08 09 10 11 11 10 9 8 7 6 5 4 3 2 1

Kriel is represented exclusively by Ultra DJ Management – cath@ultradj.co.uk
For latest updates, equipment and reviews go to http://www.howtodvj.com

Dedication

This book is dedicated to Matt Priest,
Catherine Mackenzie and Sarah Warsop, each of
whom made this period of my work possible.

pic: Pari Naderi

CONTENTS

CONTENTS

About the author

Dr Charles Kriel (aka DVJ Kriel) was born to an American circus family, and spent his childhood traveling North America in make-up and big shoes.

Kriel is now an artist, producer and DVJ based in London. He is also the United Nations World Summit Awards National Expert in New Media.

In 2000, Kriel was appointed Artist-in-Residence at BBC Radio 1, collaborating with Pete Tong to create the world's first weekly VJ stream for the web. While there, he created the first live nationally telecast DVJ mix for BBC Three / BBCi as part of Glastonbury 2004, the same year he became the first Pioneer ProDVJ. The *NME* named him the world's leading VJ; he has featured in *DJ Mag*'s Top 20, and was recently billed "the world's first superstar VJ" by *The Times*.

As a DVJ, he cut his teeth on the UK's major dance festivals, and these days you'll find him driving dirty dancefloors from Detroit to Ibiza to St Petersburg. As a visual artist, Kriel's work has included installations and performances at the Venice Biennale, Tate Britain, the Royal Festival Hall and at the gallery of Tomato Design. As a composer/producer, his works have seen premieres at London's Queen Elizabeth Hall, the Royal Opera House and Ars Electronica.

Kriel is also Senior Lecturer in Digital Media, DASS, London Metropolitan University, where he is researching his second book.

Thanks y'all!

There are many people to thank after any long project, many who should receive thanks but often don't – I hope I've included everyone:

At Pioneer: Mark Drury, Ian Jordan, Gemma Martin and the inimitable Jason Pook.

At BBC Radio 1: Sarah Martin, Andy Parfitt, Jason Carter, James Wood, Martyn Davies and, of course, Pete Tong, who let the world know about my work.

Particular props go out to (Gen.) Duncan Dick at *Mixmag* – thanks for all the beers and moral support, my friend.

Without the continued attention of Russell Deeks and Chris Kempster at *iDJ*, much of what I now do would not be possible. David Eserin, who published my first short introduction to DVJing in *DJ Mag*, deserves

special acknowledgement. And *Night Magazine's* Alex Eyre, who first sent Jason P down to my studio with the DVJ-X1 prototypes, deserves a big hug.

James Robertson and Martin Carvell at *MIM* gave special attention to this volume.

And Rich Belson is the best goddamned assistant anyone ever had.

Finally, I must express enormous appreciation and big love to Catherine Mackenzie and Brian Merchant at UltraDJ Management – my good friends always.

Front Cover Picture: Pari Naderi
Editor: James Robertson
Sub Editor: Lindsey Mann
Commissioning Editor: Catharine Steers
Photo comping: munnoo@dial.pipex.com
Cover Design: Ben van Dyke

seven tutorials

seven movie clips

HOW TO DVJ

A DIGITAL DJ MASTERCLASS
By CHARLES KRIEL

software demos

The "How to DVJ" video files

WELCOME TO THE "HOW TO DVJ – THE DIGITAL DJ MASTERCLASS" DVD. I'LL COVER A RANGE OF MATERIAL HERE, FROM CONNECTING YOUR DVJ-X1S TO YOUR MIXERS, THROUGH BASIC OPERATION OF THE DVJ-X1, AND ON TO SOME STRAIGHTFORWARD DIGITAL SCRATCH TECHNIQUES. THE MOST IMPORTANT THING I'LL SHOW YOU IS HOW TO USE YOUR PRE-MADE VIDEO MATERIAL TO MAKE A DVJ-STYLE MUSIC VIDEO IN FIFTEEN MINUTES. LET'S GET STARTED.

THE TUTORIAL VIDEO CHAPTERS

1. **Hooking up**
2. **The DVJ-X1**
3. **VJ mixers**
4. **Getting the right videos**
5. **The kit you'll need**
6. **How to make your own music video**
7. **Live in the mix**
8. **Digital scratch**

THE DVJ CONTENT VIDEO FILES

I've included about 20 minutes worth of mute video content on the DVD to get you started creating your own DVD music videos. For me, it's like a travelogue, since the selection of files comes from a range of videos I've produced over the past five years. Here are the videos, how I made them, and the small stories behind them.

Clouds and skies

These look like I found some old 8 mm film of skies in a rubbish bin somewhere, but they're actually 3D computer models. I used a program called Bryce. Bryce was once many computer artists' favorite program for creating landscapes, and many of us still use it. Lately, the program hasn't received the support it deserves from its owners, but even in older versions, it's a work of genius programming and intuitive interface. The film effects were created by bringing the rendered videos into Adobe After Effects and applying an "old film" plug-in effect. This not only simulates old film stocks, it also introduces granularity, scratches, frame jumps, and a variety of artefacts.

Thunderclub fractals

This strip of video moving from right to left was created from a series of still images I shot while touring with BBC Radio 1. When I was asked to be artist-in-residence at Radio 1, the idea was that I would tour with them and try to capture the BBC's live events in photos. For the first few weeks I photographed the bands and DJs, but bored of it quickly. It became obvious that everything on stage was a pose, even the sincerity, and the real passion was out in the audience; I turned my camera on the crowd.

The glowing effect in these videos was created using a variety of plug-in filters in *After Effects*, including a few that were designed for *Photoshop*. The movement from right to left was created in the *After Effects* timeline. The fractal effect in the center of the strip is also from a plug-in.

Deborah's flowers

While producing video for a BBC Three / BBCi broadcast of Glastonbury, I was scrambling to quickly create enough material for three nights of broadcast. My good friend Deborah Aschheim, an artist based in Los Angeles, was visiting at the time and offered to shoot some video for me. After a day at Kew Gardens, Deborah came back with this material. Brilliant! It was one of her first encounters with the creative use of video – I think she's hooked now.

After editing the video, I ran it through a bespoke video server and effects unit called *The Hippotizer*. This was an excellent alternative to rendering effects in *After Effects* (time was tight!) – I just plugged the camera in the front of the Hippo and recorded the output. Live, real-time acid effects.

Malaysian mirror

This is one of the more simple pieces of video I have made. Hiring a guide in Malaysia, I asked them to drive me to one of the world's principal Hindu temples outside Kuala Lumpur. On the way I was overwhelmed with a memory of driving along the Alabama Gulf Coast, where I spent much of my teens. The countryside was so similar! While shooting video out the back window of the car, I experimented a bit with the camera's "mirror" effect. This is the result – straight, no edits.

Berlin 360

The Love Parade in Berlin – there's nothing like it on the planet: a gazillion people, dozens of floats and sound systems, and the best DJs in the world. I shot this standing on top of the BBC Radio 1 float. Again, I used an "old film" effect in *After Effects* and used the same program to create the "letter box" aspect ratio effect.

"Dancer A"

The dancer in this video was shot at a UK Garage gig in central London, where The Dreem Teem were doing a broadcast. The DJ shots are from the BBC's

Maida Vale studios. The late John Peel had invited a group of DMC champions to mix it up on his show. As exciting as it was to watch the DJs, I was overwhelmed by the privilege of being in the studio with John. He was, and is still, a legend.

The effects on these videos were created using a stack of filters in *After Effects*. This was an early visual version of the effects setup I developed to create the look in the next video clip. It took nearly three years of twiddling to develop – the version you see here came about eight months into the process.

Deborah's flowers

Lights & lasers

More than any other set of video clips, this short group of edits came to define the look of my work at BBC Radio 1. The shots are from a variety of gigs – an outdoor stage and lasers from Homelands, the DJ booth and club lights from a gig in Ayia Napa, indoor laser lights from a gig in Paris. The effects are from *After Effects*, but the movement of the lights is strictly camera work, zooming to the 130 bpm rhythm of most house music.

THE DEMO SOFTWARE

I figure you have enough kit to collect and pay for to get your DVJ studio up and running, so to that end I've gathered together some of my favorite specialist software as demos. Enjoy!

Synthetik Studio Artist 3.5 demo (Mac)

An automatic video painting program that creates complex, painterly CGI video effects from any video file.

Flowmotion 2.8 (Mac and PC)

A VJ video trigger program that can fire off, loop, scratch, sequence and sync video clips using MIDI and other controllers.

GRID Pro

A real-time VJ media grid that allows you to access up to 800 media files using a keyboard, mouse or MIDI instrument.

Pioneer DJS (PC)

Virtual DJ software using MP3s which automatically beatmatches any DJ set, and simulates dual CDJs, a single DJ mixer and a host of effects.

1. GETTING STARTED

ON THE DECKS

A FEW YEARS AGO, PIONEER RELEASED THE CDJ-1000, AND DJS BEGAN TOYING WITH THE IDEA OF PLAYING CDS. ERM, WEDDING DJS, THAT IS.

In sweaty hip-hop joints and on the dirty, dirty dancefloors of Europe, walking into a club armed with a box full of CDs was a death sentence for your DJ reputation. CDs may have sounded better, you might have been able to manipulate them more thoroughly than records, and they may have promised many a DJ a future without an S-shaped spine, but dropping a digital tune was seen as a direct affront to the great god Vinyl.

Roger Sanchez went further than any other DJ to change this. A former architect, and always at the cutting edge of technology, Sanchez was down with the CDJ from the word go. Storming clubs across Europe, Asia and the Americas carrying nothing more than a CD wallet, Sanchez showed the DJ world just how far you could push live audio manipulation with nothing more than three CDJs and a mixer.

Within three years, most of the dance DJs at BBC Radio 1 had gone CD, and a pair of CDJ-1000MKIIs sat on every major DJ's rider, alongside the Technics and the totty.

Today's technical innovation is the Pioneer DVJ-X1 scratch DVD player, and the reaction couldn't be more different. Sander Kleinenberg, Tiesto, James Zabiela, Kriel and Jeff Mills have all embraced DVD DJing, not to mention the likes of Jamie Cullum, Black-Eyed Peas and a host of other live acts incorporating scratch video into their acts. Free to control audiences' complete audio-visual experience, the world's top performers, known and unknown, have come face-to-face with an unheard of level of creativity.

Put simply, the DVJ-X1 lets you scratch and mix from DVDs – both audio and video. On a two-deck system,

Roger Sanchez at Turnmills

pic: Lisa Loco

you can rock from the one to the two all night long, switching from music video to music video on the big club screens. Although the music and the video come from the same disc, locked in sync, you can fade the two independently, which lets you bring in video from one track before the audio, or vice versa. Add a third deck and the latest MIDI-enabled digital mixers, and a whole world of creative possibilities open up for the DVD DJ.

Along with an unanticipated host of frustrated questions.

The first being, where am I gonna get DVDs?

Sure you can drop down the record shop for a bit of rock, but for deep grooves, you're outta luck. And what if the tunes you wanna play don't have videos? How do you mix videos? How do you record your mixes? What about international video formats? Video effects? Editing? Can you use a video from

iTunes? And how do you get a DVD out of iMovie without rendering all night long?

DJing is tough, don't doubt it. But VJing for the aurally-oriented looks like a one-way ticket to geek hell.

Not to worry – we're here to help.

From the most basic steps (where to get tunes) through making your own DVDs, to the complexities of digital scratch, How to DVJ will walk you every step of the way from your bedroom to Ibiza to half-time at the Superbowl.

We'll take you through five different experiences in five sections of this book, starting in the bedroom with **On the decks**, where you'll learn everything you need to know about cueing, looping, song structure, beatmatching and more.

In the club prepares you for all the surprises waiting when you walk through the door at your first gig – hooking up kit, getting used to the sound, defining your set, and mixing video.

DVJ master takes you from making your own music videos, through DVJ effects and creating audio and video remixes, right up to getting the most from your new DVJ-X1.

Going all the way is a short section on professional practice, where you'll learn how to hook up the gig and drive your reputation through the roof.

Once you've worked your way through the book, you'll arrive as a **DVJ Jedi Master**, where we'll initiate you into the world of digital scratch.

Finally, we'll give you a few things to think about at the end – **DVJ Zen Master**. True creativity comes from an understanding of the process, but also the

Pete Tong @ Space, Ibiza

pic: Lisa Loco

ability to think about what you're doing and understand why you do it, and how it's different from doing something else. Otherwise, you're just a tune jock.

Welcome to the user's manual for the next revolution. Time to step up to the decks with *How to DVJ*.

2. GETTING DVDS ON THE DECKS

ALL THE MIXING SKILLS IN THE WORLD CAN'T MAKE UP FOR NOT HAVING THE RIGHT DVDS, AND NEXT TO READING THE CROWD PROPERLY, BUILDING YOUR DVD LIBRARY IS BOTH THE MOST DIFFICULT AND MOST PLEASURABLE THING YOU'LL DO.

If your name is Pete Tong or Qbert, there's probably a topless Russian UPS girl ringing your bell, gagging to deliver you the video you've always been waiting for. It's true, she might also give you another dozen you'll have to use for ashtrays, but it's a small price to pay for delivery.

For the rest of us DVJ mortals, we'll be scouring the record shops, garage sales and the Net trying to put together the perfect AV DVD set. Normally, that starts with the tunes.

Getting the right videos
A quick argument for making your own music videos.

The Tunes

HOW TOP DJS GET THEIR TUNES
Local record shops
The HMVs and Virgins are great places to get the Top 40 DVDs and CDs. For more esoteric tastes, you used to have to strike out to the local DJ shop. I don't like them and never have, for much the same reason I never liked the free weight room at the gym. Although DJ shops are great places to exchange road stories or just stare in the mirror, you'll get more accurate, convenient and comprehensive information about what's hot, and you'll actually be able to buy it, by going online.

Online record shops
Pete Heller: although I suspect he spends his time in the studio producing quality stormers for the rest of us, the legendary producer and DJ claims he actually spends most of his time at his computer in online record shops. What's the advantage of these shops? You can preview

The cycle of a tune

Only the world's best DJs on the world's best turntables get their discs delivered by a bikini babe. The rest of us struggle to find those stonking tunes – struggle for over a year.

The life cycle of a dancefloor stomper is surprisingly long for such a fickle fashion-based industry. Let's take a hypothetical example: DJ Killer's *Smoothjet*, which will ultimately enter the UK singles chart at number one. *Smoothjet* was born in the Spring of Year 1 as an instrumental. Over the Summer it made the top DJ rounds in Ibiza as a precious CD-R, passed around the more posh VIP rooms by the Killer himself.

Killer really did the business, so by the Autumn closing parties you couldn't move an inch without hearing a superstar DJ caning *Smoothjet*. From there it was on to life as an Essential New Tune, at which point some forward-thinking A&R guy arrived by helicopter, babe-on-arm, contract-in-hand. All he wanted was a vocal, preferably sung by a stripper.

Several lyricists and a dozen table dancers later, a fresh set of CD-Rs got burned just in time for the Miami Winter Music Conference in the Spring of Year 2. This time Killer was chillin' by the pool while his pluggers made the Miami rounds, passing out the new vocal version to every playa in the Top 100 DJs list.

If a tune's a winner at the WMC, it will typically go on to become the Summer's Big Tune in Ibiza – erm, a year after it was the Summer's Big Tune in Ibiza. An imminent Autumn Year 2 release takes it into the Buzz Charts, hopefully at number one. And you, lucky DVJ, can buy it in October – a year and a half after Killer first handed it over the mirror to DJ Superstar. Look for a special edition DVD release just in time for Christmas!

every track in the store without waiting in line or wading through a roomful of I-wanna-be-a-DJ-but-I'm-stuck-in-a-Nick-Hornby-novel attitudes. And you can buy the track and burn it instantly.

Pluggers

Record pluggers are hired by labels to promote their tunes to top DJs and radio programmers. You won't even meet one of these guys until you've got your own afternoon drive slot in Atlanta – by then you'll hate 'em anyway.

Subscription DJ services

Promo Only, CD Pool, Mixmash and a few others worldwide provide a monthly subscription service tailored for DJs. You'll get about 20 tunes on a CD or 35 videos on a DVD. Depending on the level of your taste, you'll find about 5–35% of them playable.

Illegal MP3 downloads

The operative word is illegal. You'll have to wrestle with your own conscience here. If there is a benefit to illegal downloads, particularly via Soulseek and other dance-led peer-to-peer file-sharing systems, it is this: the abundance of unreleased and out-of-print tracks. Many producers, pluggers and even young bands also make it a habit to upload their latest promo tracks as a way of getting them into the hands of the maximum number of DJs. Either that or there are a lot of bitter and twisted record label employees in the world. Either way, DJs win. I'm not sure about the artists though, and I hope the record industry stops trying to sue grannies or throw DJs in jail, and instead sorts out a way for us to pay a subscription for unlimited quality online file-sharing services ... now!

Legal MP3 downloads

2004 was the year independent dance labels got smart and started selling MP3s online, specifically for DJs. Plus, they started giving log-ins and passwords to top DJs so they could download the latest promos free. Whether you're buying from one of the dozens of MP3 shops online, or blagging a password by saying you're Paul Oakenfold's sister, there are several advantages to legal downloads over illegal:

- Guaranteed high quality MP3s and AIFFs
- No viruses
- No RIAA lawsuits
- No jail term.

WHICH TUNES TO DOWNLOAD?

Your first instinct will scream "all of them!" A couple of terabytes of hard drive later, and you'll own the most unmanageable collection on the planet. My vinyl and CD collection is so disorganized, I find it easier to go out and buy a second copy of a tune, rather than find it on the shelf. Multiply that times 1000 and you've got your typical MP3 collection. Be a little selective.

Here's what I go for digitally:

- Classics – the ones that sound fresh after a decade of play.

- Forgotten gems – an unknown or mislaid track from the 70s can drop into a contemporary set like a diamond. Just remember, production values and therefore the quality and punch of the basslines were held to a different standard 30 years ago. Unless you're Gilles Peterson, proceed with caution.

- Recurrents – they were great three years ago and, played sparingly, they make an otherwise unremarkable set legendary.

- Rare mashups and unusual remixes – why would you want to sound like everyone else? Just be sure to get the original, too, in case it becomes a classic.

- What you like – unless you'd prefer to spend your life playing what other people like.

Taking the pulse

The internet is your greatest resource for what's happening outside your bedroom or resident club. Here are some favorite sites:

- BBC Radio 1 (www.bbc.co.uk/radio1) – Pete Tong, Westwood, Judge Jules and the residents. Checking their playlists should be a weekly ritual.

- *DJ Magazine* (www.djmag.com) and *iDJ Magazine* (www.i-dj.co.uk) – monthly and bi-monthly charts

of House, Hip-Hop, Breakbeat, Garage, Techno, Leftfield, Drum and Bass and more. Also check the occasional guest DJ slots. Or you could go one better and subscribe to the mags.

- DJ websites – who are your biggest influences? Check their charts to find new tunes you'll love.

- Producers and remixers – check their websites for regular updates on their releases. The Scumfrog (www.thescumfrog.com) not only produces some

Legal MP3 downloads

I would never condone illegal MP3 downloads. Artists get short changed enough by their record labels without us encouraging the CEOs to give even less money to my heroes. But when forward-thinking dance labels like Subliminal started offering their wares via MP3, the local DJ shops started closing their doors and moving on to the internet frontier, helping create a new category of tune – the legal MP3 dance download.

This couldn't be better for the DJ. You can preview the tunes online, download them for a dollar, burn 'em to disc and hit the club all in the same night. Here are a few of my favorites:

3beatdigital.com • anjunadigital.com • audiojelly.com
beatport.com • bleep.com • digibag.com
djdownload.com • playittonight.com • trackitdown.net
traxsource.com • westendrecords.com • xpressbeats.com

of my favorite tunes, he also has one of my favorite websites, providing regular free and legal downloads of his latest tracks.

- DVJ Kriel (www.kriel.tv) – check my website for regular video and clubland chart updates.

- Resources – possibly the most comprehensive list of contacts for the dance music industry is at www.mim.dj. Ranging from promoters, VJs, nightclub owners right through to equipment manufacturers and record labels.

The videos

HOW TO GET THOSE DVDS

I've already pointed you to HMV, Virgin and the subscription services for tunes – they're also great for DVDs, if you want to play Britney or Beyonce. Anything else, and the story gets complex. House and Hip-Hop jocks who like their tunes deep know there are

few videos to match their tastes. And although there are plenty of pre-mixed mute VJ DVDs around, if you wanted your sets to look like glorified screensavers, you wouldn't have bought a pair of DVD decks.

The unfortunate truth is, you'll need to make your own DVDs if:

- You play House, Trance, underground Hip-Hop or any other discerning dance style

- You want to add screen-based DVJ flair to your sets

- You want to drop AV samples into your sets (believe me, you want to)

- You're tired of watching that FBI warning every time you go to cue up a tune.

The good news

It's really easy – just check out the **DVJ Master** section later in the book.

3. CHOOSING THE RIGHT KIT

ON THE DECKS

HOOKING UP WITH THE RIGHT KIT IS ESSENTIAL. ALTHOUGH A GREAT VIOLINIST CAN GET A SWEET SOUND FROM A STRETCHED CAT, THE ONE THING THEY COULDN'T DO IS LEARN TO BE A GREAT VIOLINIST WITHOUT PRACTICING ON A PROPER INSTRUMENT. DODGY KIT IS KINDA LIKE THE BACON AROUND THE FILET, IT HIDES A LOT OF SINS. LEARNING TO MIX ON YOUR HOME STEREO THROUGH A CHEAP TWO-CHANNEL MIXER IS FINE UNTIL THE DAY YOU PLUG INTO A PAIR OF DECENT MONITORS WITH A DJM-800. WITH THE KIND OF CLARITY GREAT KIT GIVES YOU, YOUR SINS ARE EXPOSED AND YOU FIND YOURSELF LEARNING TO MIX ALL OVER AGAIN.

While you might not be able to buy the best kit right off, I'll point you toward the industry standards, and some great home kit that'll get you started.

DVD decks

Pioneer DVJ-1000

A second generation DVD player, based on the DVJ-X1 but 20% smaller and 40% cheaper! Virtually all the same features but now including MP3 compatibility (with an intuitive navigator), a brighter and larger WAVE display, studio quality 96 kHz/24-bit audio and several new innovations like 4X hyper jog mode for track searching and a first ever "back and forth" loop button. It's a globetrotter too, playing both NTSC and PAL DVDs using a built-in standards converter.

Pioneer DVJ-X1

There really are only two choices for DVD decks, and that's the DVJ-1000 and the DVJ-X1. DVJs need from their DVD decks precisely what they would expect from their turntables – instant start, rock-solid performance, scratch without delay, pitch-bending, digital sound, cue points and no glitches. Although there are cheaper decks on the market, most lack any or all of these features. This is the one place to splash out your cash – it's worth it.

Pioneer's DVJ-X1

Numark DVD01 Dual DVD Player

Awesomely cheaper than the DVJ-X1, the DVD01 gives potential DVJs two decks in a rack-mount configuration for less than the cost of one – but not without a price. If you're at all interested in beat-matching, creating seamless loops or scratching, you can forget about it. While the decks will cue up a DVD, they won't start instantly, producing a noticeable pause between pressing play and play actually happening. There is a gap when you create a loop, meaning you'll have a moment of silence and black before your selection loops around to the front. And there's no scratch facility. I also used a pair of these on a North American gig recently, and found they wouldn't play long PAL tracks from beginning to end without stuttering.

However, if you're a mobile jock playing mostly corporate, wedding or high school dance parties, these decks may get you on the DVJ ladder at a price you can afford. Just be prepared to upgrade at a later date.

CD decks

CDJ-1000MK3

This is the industry standard for CD decks. You'll find them installed in nearly every major club on the planet. And the few clubs not rocking the CDJ-1000MK3 still have MK2s or the original CDJ-1000. There are lots of pretty CD decks out there, no doubt.

CDJ-1000MK3

However, I think it's best to learn on installation-standard kit – that is, the kit that will be in the club.

Whether you get yourself a pair of MK3s, or a used pair of CDJ-1000MK2s, I suggest sticking to the top of the line.

Technics SLDZ1200

Although Technics dominated the DJ scene with their vinyl decks, they've been slow off the CD mark. The SLDZ1200 was a fine start, however. Featuring SD card memory, a sampler, looping and MP3 play, the SLDZ1200 combines many of the features of the higher end of Pioneer's kit with lower grade scratch DJ decks. And it comes with a spinning turntable to better replicate that vinyl feel when scratching from CDs.

If there's a problem, though, it's that these beautiful decks have been largely ignored by DJs and club owners alike. Whether for their size (they're big!), they're synthy-sounding scratch algorithms or the camp look of silver, I'm not sure. Just don't expect to encounter them in a club installation.

[James Zabiela]
DVJing for me is just another way to have more fun behind the decks. Another element to control and manipulate and another dimension to take the crowd into.

Audio monitors

Audio monitors are a matter of taste. The one thing you want to be sure of is that you get the cleanest, flattest, fattest sound available.

Lots of DVJs start off mixing on their home stereo. This only works for a while. First, home kit isn't capable of the volume you'll be dealing with in the club. Although you should always protect your hearing, rehearsing loud is sometimes a necessity; you'll need a monitor/amp system that can handle the load.

You'll also want as flat an amp as possible. This means an amp with minimal spikes in the treble, bass or mid-range.

I use Event's Studio Precision Direct Field Monitor System (www.event1.com) in my studio. In the active versions, these are powered monitors, so there's no need to go through the hassle of matching an amp to the speakers. They're powerful, full of bass and have an amazingly flat field. And they're pretty damn cheap!

Other brands to look at include Yamaha, Alesis and about a billion others. Just have a good look around, do your internet research and listen before you buy.

One thing to keep in mind with near-field monitors is that they have a sweet spot – that is, one spot where you have to stand in the crossfield to hear everything. Move outside that sweet spot and you'll lose parts of the sound – bass or treble will disappear and you'll have no idea what happened. Nonetheless, for rehearsing, remixing and producing, I don't think you can beat the Event Studio Precision series for the price.

Video monitors

Old skool VJs used to go for the BBC look in video monitors at gigs – the kind of big, deep, tube-based Sonys you would typically find in a television studio. Backstage at Creamfields, Gatecrasher or one of the other giant rave festivals, he'd find a VJ sitting behind a bank of monitors you'd need a white van to haul from gig to gig.

There were two advantages to this approach: first, the video could be perfectly tuned and what you saw on the monitor was exactly what was on the tape; second, you could charge the promoter a bundle for the rental of the kit.

New school DVJs take a different approach. While you might want the best monitor in your studio (and here I use a Pioneer flat-screen), on the road you need to travel light.

I use a unique setup for monitoring in the club, and it might work for you: my handheld video camera. Just attach the preview monitor output from the mixer to the analog input of your camera, *et voila*, instant multi-tasking, ultra-portable, super-cheap video monitor.

Cameras

There are a range of cameras available, in half a dozen different formats. From mini-DV cameras through to the new generation of HD cameras with hard drives, nearly all are appropriate for the DVJ. I prefer the Canon XM series (or GL in North America), particularly for its balance of features, quality and friendliness. The camera you choose will really depend on your budget and taste.

I recommend you look for these features:

- S-Video in and out
- Composite in and out
- DV in and out
- Digital and optical zoom
- Digital and optical stabilization (for image shake).

Some features to look out for on your camera

1. Digital video (i.link/Firewire)

2. USB jack for still images

3. Component Out jack

4. A/V Out jack for connecting to the TV or analog recorder

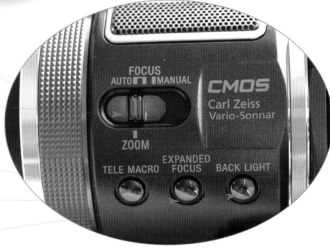

5. Auto focus or manual

6. Macro for close-up photography

7. Use this to get a better focus

8. Very useful when shooting into a bright background

DJ mixers

DJM-1000

This is the ultimate mixer, full stop. From the point of view of broadcasters and venues, the Pioneer DJM-1000 is a robust, six-channel, installation grade mixer with super-quiet digital audio and intuitive design. All great, but for the DVJ, the DJM-1000 is quite simply a paradigm shift. Internally, you can send the sound a dozen different ways without any distortion. Two effects – sends and crossfader-effector control – open up creative territory unseen in any previous DJ mixer. And the feel of the custom-made sliders is just plain fun – unless you opt for the rotary pots! It's pricey for the bedroom DVJ, but it's also sure to become the new clubland standard. Best of all, its digital connections can take digital outs from the DVJ-X1, as well as send and return digitally to the EFX-1000. And the MIDI output connects your crossfader to the T-bar on your video mixer. I'm in love!

DJM-800

The DJM-1000 is the bomb, but at a price few can afford. More in line with most DJs, whether superstars or bedroom heroes, is the Pioneer DJM-800. You'll sacrifice a couple of inputs, as this is a four-channel mixer, but in return you get two banks of effects (including filter sweeps) and you'll save a grand. The features of the DJM-800 are extraordinary, particularly its MIDI capability. Each knob and button on the DJM-800 sends MIDI controller data.

DJM-800

This can be used to control a number of external devices, from effects units to MIDI software on laptops. Most important for me, however, is its ability to send MIDI signal to video applications and hardware.

The DJM-800 can be used to control the Edirol V4 video mixer. Each knob and fader can tweak a different effect. The same is true for MIDI-enabled VJ software. How does this translate in the real world? If, for instance, you set up your equipment so the bass EQ knob on channel one of the DJM-800 controls the coloration of the video on bus A of the V4, then when you bring in the bass on channel one, not only will the sound change, but the video could become, say, more blue. Each time. This is a real paradigm shift for DVJs and cannot be underestimated. As the DJM-600 was the installation mixer of choice for the analog generation, so will the 800 be for digital AV gurus. A real winner.

DJM-600

The DJM-600 was the rock-solid workhorse mixer of the best clubs in the world for a generation of DJs. More DJs specified this four-channel mixer in their tech rider than any on the market. With clean sound, a great build, effects sends and a range of setup options, the Pioneer DJM-600 and DJM-600S (S for silver) are still workin' for House and Scratch DVJs alike. If the DJM-1000 is the Hummer of DJ mixers, and the 800 a Maserati, then the 600 is still a bad-ass Alpha Romeo.

Audio effects

EFX-1000 and EFX-500

The Pioneer EFX-1000 and 500 are truly a DVJ's best friend. A pair of kickin' DJ effectors, the EFX series are the basis of nearly every digital scratch technique on the block. The 1000 drops in at nearly twice the price of the 500, but the sound quality, range of effects and total digital integration with the DJM-1000 make it well worth the price for anyone able to step up with the cash.

The EFX-1000 is an all-digital effector with 96 kHz audio, a simple but intuitive interface and every effect the most demanding DVJ could ask for, including vocoder, pitched echo and a humanizer.

The EFX-500 may lack a few of the effects and the pure digital throughput of the 1000, but it still formed the basis of James Zabiela's, Erick Morillo's and Dan Tait's technique for years. It's a brilliant professional effects unit at an entry-level price, and still goes out in my record bag every gig I play.

If you're in the market for a DJ mixer as well, however, consider the DJM-800, with two banks of onboard effects. Just make sure the club you're playing this weekend has one or you'll be left high and dry in the mix.

Korg Kaoss Pad Entrancer

I love none of my kit as much as I love my Kaoss Pad Entrancer (except maybe the DVJs). An audio and video effects unit rolled into one, this is top kit to die for. More below under 'Video effects'.

Video mixers

Edirol V4

On top of a DJ mixer, you're going to need a separate VJ mixer to switch and fade from video to video – that is, until someone invents the perfect DVJ mixer.

While VJs have been lugging around big Panasonic industrial-grade mixers for yonks, the last two years have seen the V4 replace the lunky Panasonics in almost every VJ's rig. Small, light and built for the club, the V4 takes four inputs, gives two outputs plus preview, has over 200 transition effects, features MIDI In/Out and Through, and lets you beatmatch effects. Even better, with MIDI Out now specified on many new digital mixers (particularly the DJM-1000 and DJM-800), you can control the V4 from your DJ mixer. Although the box can be a bit delicate at times, and the menu system is both ugly and wonky, you really can't beat the V4 as an all-round mixer on price, performance, ease-of-use and club compatibility. A winner.

Korg KrossFOUR

The KrossFOUR from Korg is a new entry into the video mixing arena. Built to match Korg's Kaoss Pad Entrancer video effects unit, the KrossFOUR on the surface appears to be a cut-down V4, with far fewer transition effects. Truth be told, most transition effects for video are crap, and you'll never use them. I think the KrossFOUR's four fader effects are more than enough. Recent price reductions mean many DVJ budgets have room for both the KrossFOUR and the Entrancer: if that includes

yours, you'll be laughing your way to DVJ heaven. Highly recommended.

Pioneer VSW-1

The VSW-1 is a transitional product introduced by Pioneer to match the DVJ-X1. The VSW-1 does one thing – cut from A to B – but it does it with style. All the component, composite and S-video leads are there, putting it a technical jump ahead of most mixers. It will also automatically cut when you activate the upfaders on any Pioneer mixer with fader start. And it's small, light and robust – smaller than either the V4 or KrossFOUR – brilliant for gigs. That said, cuts-only video switching is limiting. This one's great for gigs, but you'll need another mixer in your studio.

Korg KrossFOUR – front view

Video effects

Korg Kaoss Pad Entrancer

The Kaoss Pad Entrancer is really the only game in town when it comes to video effects for the DVJ. The Entrancer is set up with an XY pad you run your finger across to change the parameters of more than 100 video effects. It also has a short video sampler which lets you grab clips and scratch them back and forth. The build quality is solid, and the unit is designed to match the Korg KrossFOUR in look and feel. I love both the expensive looking time-slicing and the crunchier bit-chewed low rez effects. The XY pad is a truly expressive instrument, and the look of the machine itself is ace. In the realm of video effects, it's the Entrancer or nothing.

Nearly nothing. For the more adventurous of you, a laptop with Vidvox Grid Pro is another route, if you're a computer nut. But I suggest sticking with the Entrancer for pure effects.

VJ software

FOR PERFORMANCE AND PRODUCTION

VJ software is an essential tool for both DVJ and VJ alike. Unlike standard video editing software, VJ software is designed with live performance in mind. With normal video editing software, the standard process is to import video from your video camera, create your edits and effects, then export the movie, either to tape or DVD.

With VJ software, video is imported into the computer using a standard video editing program, as above. It is then broken down into short clips – anywhere from one second to a couple of minutes. Those video clips, or files, are then imported into the VJ software, and prepared in much the same way a keyboardist prepares samples for stage performance. Each video clip is assigned to a different key on the computer keyboard (or external MIDI keyboard), and the clips are "played" live, the way you would play the piano. In the club, the computer's video output is routed one of several different ways to projectors or plasma screens. But VJ software can also be used in the studio as a production tool and, in this case, the video output is either routed to a DVD or tape recorder.

VJ software is generally inexpensive, fairly simple to use (after a bit of grappling with the interface) and stable enough for live shows.

Video and VJ editing software

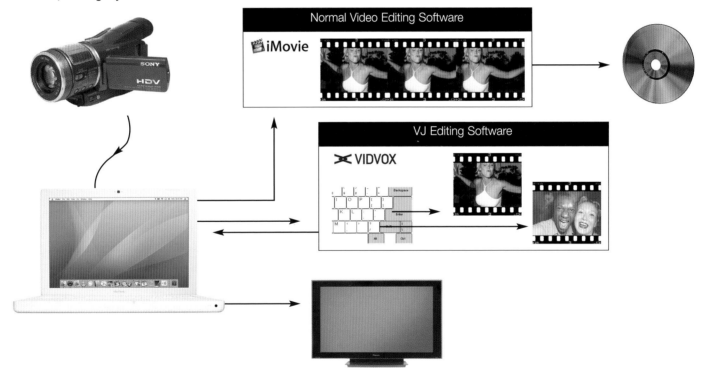

Normal Video Editing Software

iMovie

VJ Editing Software

VIDVOX

Vidvox Grid Pro (Mac), **Vidvox Grid 2.0** (Windows)
The boys at Vidvox have been producing the best VJ software on the market since video artist Johnny Dekam set up the company in 1999. Grid Pro is an amazingly fast and simple program for firing silent video clips from either a MIDI keyboard or via the keyboard of your computer. Think of it as a live performance video editor. Video can be output straight to screens or, more useful for DVJs, straight to a DVD recorder, letting you make great music videos for your favorite tracks in no time. Grid Pro, introduced in 2005, adds three layers of effects, audio analysis, text generation, recording, MP3 playback, live input, sequencing and gesture control. Nothing on the market touches Grid Pro. As for reliability, I've used Vidvox software on the road with the BBC for five years, with nary a glitch or crash the whole way. Outstanding software. I wouldn't use anything else on my Mac.

VJ hardware
PERFORMANCE SAMPLERS
Korg Kaptivator

Korg's Kaptivator is a live performance-oriented video sampler. Heaving with effects, the Kaptivator provides rotary, ribbon and crossfader controllers to manipulate up to 800 video clips sampled onto

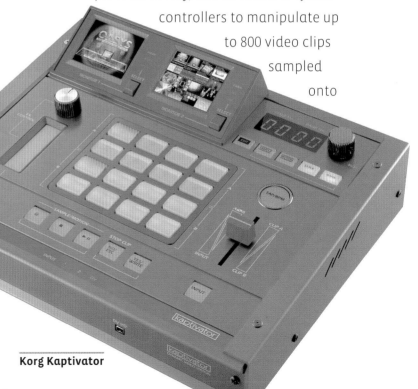

Korg Kaptivator

its 40 gig hard drive. Although the Korg costs more than, say, Resolume and a kicking laptop, there are real advantages to this piece of hardware. First, its hardware. That means less crashing at gigs, robust design and a smaller footprint when you're trying to squeeze into a tiny DJ booth. Second, it's a Korg. Korg make musical instruments, and this video sampler is built with the kind of friendly and intuitive interface you would expect from a sampler or drum machine – read, it looks a lot like those amazing Akai drum samplers. Finally, if you can only afford one piece of VJ production/performance kit to accompany your DVJs and mixer, either because your wallet or your suitcase is too small, then this is the one. It does the job standalone, and works straight out of the box. If you're gigging regularly and gigging hard, I recommend this brilliant piece of kit.

Recorders – DVD

There's no small shortage of DVD recorders on the market these days. Where once you paid thousands for the privilege of recording to DVD, now you can buy a DVD recorder at your local Piggly Wiggly, along with the cheese.

As for which to choose, think compatibility and practicality. I always author on the same brand of kit that I play back on – which is to say, use the same brand of recorder as you do player. On the practicality front, DVD recorders with hard drives are getting cheaper. They'll save you a fortune in wasted media, and with the size of HDs these days, you'll have plenty of room for all your DVD music videos, plus a full season of *Deadwood*.

ON THE DECKS

ONCE YOU'VE ASSEMBLED YOUR DVJ ARSENAL, YOU'LL ALSO HAVE A CONFUSING BIRD'S NEST OF CABLES AND CONNECTORS. IN MANY *MUSIC* STUDIOS, YOU'LL FIND YOURSELF SETTING UP ONCE AND FORGETTING ABOUT IT. IN THE *VIDEO* STUDIO, YOU'RE MORE LIKELY TO SPEND YOUR FAIR SHARE OF TIME CRAWLING AROUND ON YOUR HANDS AND KNEES REWIRING ON A DAY-BY-DAY BASIS. VIDEO PATCH BAYS (LIKE AUDIO PATCH BAYS) ARE A POTENTIAL SOLUTION, BUT PROHIBITIVELY EXPENSIVE.

You'll also find yourself crawling around in the dark once you hit the venue. This chapter is here to help you make sure you put the right plug in the right hole, even when you're under the DJ desk.

This is also the chapter with lots of acronyms and abbreviations. Chances are after you've read it, you'll forget a lot of it. But what you need to remember is crucial (like voltages).

The essentials

- Buy the best cables you can afford – most will be phono plug cables or power cords

- Connect the cables with the equipment turned off

- Turn your speakers on last and off first

- Mix and match equipment from North America and the rest of the world carefully

- And fill the gas tank before you bring the car home, son. The rest is detail.

110/220 – the most important section in this book!

Beware the "mains" (that's AC to North Americans).

North America, parts of South America, Japan, Saudi Arabia, Madagascar, Sierra Leone and Singapore use 110–127 volt power systems. Everywhere else uses 220–240 volts.

WARNING TO JET-SET DVJ SUPERSTARS: plug the wrong one in the wrong place and you'll be "firing up" your decks, for real! I've seen more than one DJ who shoulda known better ...

Keep in mind, too, that cheap plug converters (from US plugs to Euro plugs, for example) CAN'T CONVERT VOLTAGE! They only make the physical parts fit. As a rule, voltage converters cost a minimum of 50 of your local currency in euros, dollars and pounds – or a gazillion yen. If it cost less, have a fire extinguisher ready when you turn your decks on.

Cables

There's only one rule here: buy the best you can afford. No point running a $15,000 AV system through $2 cables.

Powering up

You can blow things up by turning them on in the wrong order, so I'll repeat this. Turn your kit on starting at the sound source – your decks – and working your way to the amps, with all the faders down. Turn your equipment off in reverse order.

Audio
SOME TERMS – dB, BALANCED AND UNBALANCED

These all have to do with the type of audio put out by a piece of equipment. Electro boffins like to divide kit into pro and everything else. These are the guys who'll pack up your kit after you've performed. Don't be intimidated; pros use whatever works.

Balanced

<u>Pro</u>: balanced plugs and cables usually have three wires. They let you make long runs of cable, and are less subject to noise and interference from other equipment. The connectors are usually XLRs, which look like a microphone plug.

Unbalanced

Not pro: unbalanced connections are usually phono, or RCA cables and plugs. They're not very reliable, so it's best to keep them short, or go for the shielded cables, which are also the standard connectors on DJ equipment. Go figure ...

dB

dB is short for decibels, which is a way of measuring voltage, volume and a few other things.

DVJs who are hooking up their kit need to concern themselves with two types:

*   −10 dB – unbalanced, semi-pro gear. Has a lower voltage (volume) level

*   +4 dB – balanced, pro gear. Higher level.

You need to worry about this if your gear supports both balanced and unbalanced audio leads.

Generally, don't mix and match.

SEND AND RETURN

This sends an audio signal from your mixer to an external sound unit, usually an effector (effects unit). It saves running a signal straight from the output of your deck into the effects unit and then into the mixer. It's usually only a feature of top-end gear, but pays for itself when you don't have to buy two effectors, one for each deck.

DIGITAL OUTPUTS

DVJs, some CDJs and quite a few DJ mixers have phono plug digital outputs. You can use a standard phono cable but, again, this is like sticking a $5 filter on a $5000 camera. I suggest buying proper digital cables.

CONTROL CABLES

CDJs, DVJs and mixers have control cable connectors which automatically start the decks when you push

Signal flow of send and return

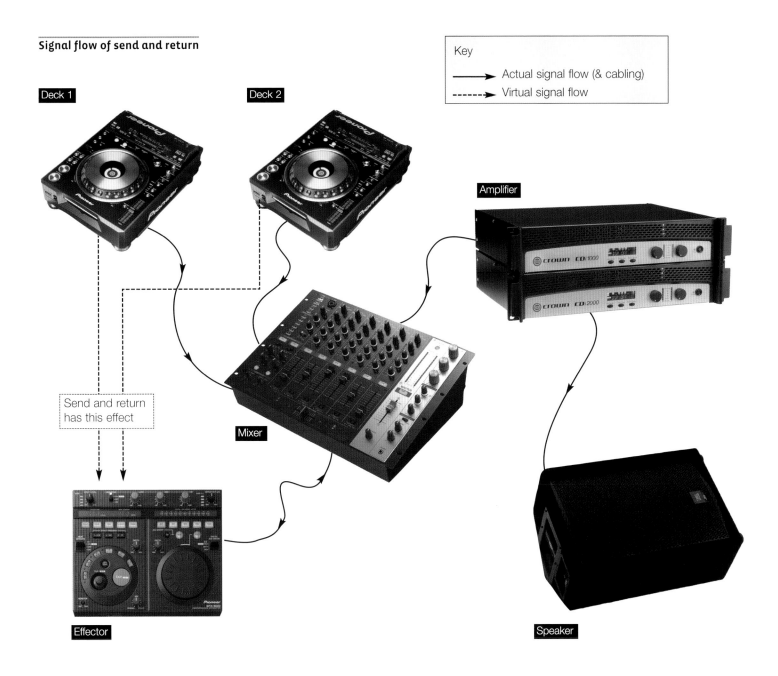

Key

→ Actual signal flow (& cabling)

--→ Virtual signal flow

Deck 1

Deck 2

Amplifier

Send and return
has this effect

Mixer

Effector

Speaker

the fader up, or when the track on the other deck runs out. Seriously – have a look. Although few top-end DJs are even aware of this feature, there are some applications for DVJs – specifically, the control cable can be used to switch video. See the Pioneer VSW-1 on page 26.

Video
PAL/NTSC/SECAM

When they made video, they divided it into American, European and a combination of France and the Soviet bloc.

Here's what you really need to know.

Generally:

- NTSC = North America, South American west coast, Japan and a few other places

- PAL = most of Europe, South American east coast, parts of Africa and Australasia

- SECAM = France and the ex-Soviet bloc.

I say "generally" because there are subcategories of each, and in broadcast situations they're incompatible.

- Most video projectors everywhere accept PAL and NTSC, as does most every piece of European video equipment

- The latest DVJ-oriented video mixers accept both PAL and NTSC, but not both at the same time

- SECAM is the most incompatible system

- Generally, PAL looks best.

Global broadcast formats

NTSC

PAL

SECAM

COMPOSITE, COMPONENT AND S-VIDEO

Not only are there three video formats based on political geography, there are also three video formats based on use: component (pro), composite (home) and S-Video (fakin' it).

Composite video

This is what you're used to. It's the standard for home video equipment. It only uses one cable for video, typically with a yellow RCA/phono plug on each end, and two cables for audio. It can also use a BNC plug, which locks onto the socket of the video equipment. BNC cables are shielded, which is a good thing, as it keeps stray radio frequencies (RF) from interfering with your video signal.

Video geeks will tell you composite video isn't "broadcast quality". Tell them "rubbish". I've seen a group of BBC engineers wire a DVJ-X1 into an oscilloscope using composite video, and collectively nod their heads in approval. If it's good enough for them, who are the video geeks to disagree?

Component video (the three-cable kind)

Video signal can be divided into several components in several ways, including chrominance (color) and luminance (brightness), red, blue, green, yellow, sync, bathtub, etc. You're dealing with a component video connection when you have a bundle of three cables with BNC connectors on the ends, or when you use a European SCART cable (possibly invented by the French).

Unless you're DVJing on MTV or in an arena with more than a quarter million people, you'll rarely use component. But the video geeks swear by it ...

S-Video

AKA Y/C. Technically, it's component video, as it divides the chrominance and the luminance. You can tell you're using it when you've plugged the thick video cable with four pins into your decks. It's better than composite, but not as good as "three-cable component". Even cheap club-based video mixers accept this kind of video, as do most good camcorders. But a word of warning – the cable design for S-Video is more hassle than it's worth. Never leave the house without a spare, as S-Video cables break far too easily. I suggest you never use it.

CONNECTING YOUR VIDEO KIT

The same rules apply here as with audio:

- Connect the kit with all the power off.

- Power up from the source to the output – DVJ to projector. Power down in reverse.

5. FIRING IT UP: UNDERSTANDING TUNEZ

ON THE DECKS

GUITAR PLAYERS HAVE TO UNDERSTAND THE STRINGS AND FRET SYSTEMS. DRUMMERS HAVE TO LEARN THE RUDIMENTS. PAINTERS NEED TO KNOW THEIR OILS FROM A HOLE IN THE GROUND, AND IT'S BEEN SAID THAT A CHESS CHAMPION HOLDS 10,000 DIFFERENT BOARD PATTERNS IN THEIR HEAD.

DVJs? We've got it easy. All we need to understand are tunez – most of us are already there, but are unaware of it. We just don't know how to name the things we feel instinctively.

If there's one concept to learn it's this: song structure.

The language of music
LEARNING TO SPEAK WHAT YOU HEAR

Your beatmatching can be slicker than Sasha's, but if you don't understand song structure you will always be second rate. Master song structure and you can fake the rest.

You don't need a semiotics professor to tell you music is a language. But music videos are a language, too.

The best part of learning to DVJ is realizing how much you already know. Before you learn to speak a language, you have to understand it. And music is something you definitely understand. I'm here to teach you to speak.

DANCE RECORDS, AND ALL THE REST

For DJs, there are only two types of tracks: dance tunez, and all the rest.

Dance tracks are made specifically for DJs. Back when the earth was still cooling off, they were called 12" singles. Now I call them edits, dubs, remixes, rerubs, bootlegs, white labels and a few other choice names depending on their quality. What marks them out is length – they're longer than normal tunes, with extended intros and outros for easy mixing. Playing dance records is like long, slow lovin'.

All the other tracks were made to play on the radio or sing along to.

COUNTING TO 64

If you can count to 32, you can DJ. If you can count to 64, you're a pro.

Almost all popular dance music is in 4/4, or "four-four" time. You count the beats – **1**, 2, 3, 4, **1**, 2, 3, 4. One count of four is called a bar.

Do that eight times and you've counted to 32. A group of bars that go together usually travel in collections of eight or 16, for a count of 32 or 64.

Throw on your favorite dance tune. Notice how everything happens in little groups of 32 or 64 beats, or divisions thereof? At the end of 16 beats, you'll hear a wee extra something in the drums. At 32, you'll hear a more dramatic marker. And chances are, at 128, you'll hear the music structurally change. That's

the end of a musical section.

SONG STRUCTURE

Tunes are typically divided into sections: intro, verse, chorus, breakdown, middle 8 and outro.

They can come in any order, although there are standards. Here's what they're for:

Intro

The intro to a tune is the part that comes before you get into the song's body. On radio-oriented tunes (the ones you hear on the Top 40), that'll be the instrumental introduction before the singer does their thing. This is usually 32 beats , or eight bars long.

On a dance record, the intro will be longer and divided into a couple of sections. First will come a simple drum pattern for 16 or 32 bars (64 or 128 beats). Then comes a slightly more complex drum

Counting to 64

Bar

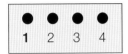

8 Bar – you can DJ

1	2	3	4	5	6	7	8
1 2 3 4	2 2 3 4	3 2 3 4	4 2 3 4	5 2 3 4	6 2 3 4	7 2 3 4	8 2 3 4

16 Bar – you're a pro

1	2	3	4	5	6	7	8	9	10	11	12	13	14	15	16
· · · ·	· · · ·	· · · ·	· · · ·	· · · ·	· · · ·	· · · ·	· · · ·	· · · ·	· · · ·	· · · ·	· · · ·	· · · ·	· · · ·	· · · ·	· · · ·

pattern, maybe with a bassline, for the same number of beats. The intro to a dance record is there to help you beatmatch with the outro or final choruses of the tune playing out on the other deck.

Beatmatching happens when you align two tunes perfectly, so they share the same beats per minute. Your first tune might be 132 bpm (beats per minute) and the second 136 bpm. If you try to mix one into the other without adjusting the speeds so they match, your perfect mix will sound more like tennis

shoes in the dryer. Adjusting the tempos to match are what the superstars do to get that long, all-night mix. And the intro is there specifically to help.

Verse
The verse is the part of the tune that tells the story, or sets up the situation. "When I was young, it seemed that life had just begun, blah, blah, blah ..." This will be repeated in total, or in a variation, several times in the song.

In a dance tune, the "verse" is usually the rolling beat, bass and high end, with a melodic twiddle. Or on a dance remix of a song, it will be a single, simple vocal phrase from the original, sampled repeatedly with effects.

Chorus

The chorus is the lament. If the verse gives you the background, the chorus tells you what is happening now. That old guy moaning about what life's like now would sing "all by myself, all by myself" until you wanted to shoot him. Kelis sings "My milkshake brings all the boys to the yard, and damn right, it's better than yours." She sings it twice, and that's eight bars or 32 beats.

Chorus – Kelis's Milkshake

The chorus is slightly more complicated than it seems. The eight-bar "phrase" actually starts on the word "milkshake", but the vocal (i.e "my") starts at the very end of the previous bar. This is referred to as an "upbeat" and is repeated throughout the eight bars. Notice the position of the word on the beat: "My" is placed towards the end of the beat, because it is an upbeat

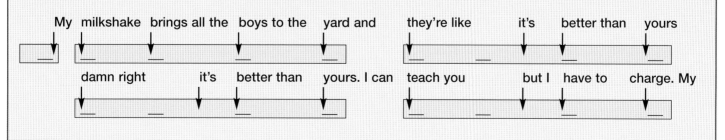

On a dance track, the "chorus" is usually a repetitive melodic hook.

Breakdown

This is the exciting bit in the middle. On dance tracks I call it the breakdown, but on other tracks it's called the break. You can always spot a break on your dad's vinyl, because there's someone playing too much guitar or sax. Breaks are usually the length of a verse and chorus combined.

A breakdown on a dance track is where a lot of the tune disappears, so the punters can put their hands in the air and anticipate the release they'll feel when the chorus returns. On cheesy trance tunes, it usually ends with a drumroll. I call it the "build", and it almost always ends with either a big accent or dropout on the four. A breakdown is anywhere from 132 to 264 beats or more, depending on the length and style of the tune. A dance track almost always has two breakdowns, the second one half the length of the first.

Middle 8

The middle 8 is the vocal part of a normal record that is different from both the verse and the chorus. On Beatles records, it's the part where John Lennon sings something that has nothing to do with the rest of the song. It's often eight bars long (32 beats), thus the name.

In a dance track, the middle 8 is usually the build at the end of the first breakdown, before the chorus kicks back in, and the punters go bonkers. Dance tracks will have two versions of the middle 8, one for each breakdown – and despite the name, they could be any length.

Outro

This is the repeat of the chorus until the tune fades out.

On a dance record, it's more like the intro, but at the end.

PUTTING IT ALL TOGETHER

You can put these bits together in any combination, although the intro is usually at the front, and the outro.... Listen to your favorite tune and figure it out. Then try comparing it to other tunes.

Here's a hint to get you started on normal tunes: dance tracks are unpredictable structurally. Sometimes the verse is the chorus, sometimes the chorus is the middle 8. You never know.

Song structure

Intro: 9 Bars ①

Mini intro: (1 Bar)

My milkshake brings all the boys to the yard: (4 Bars)

My milkshake brings all the boys to the yard: (4 Bars)

Verse : 8 Bars ②

I know you want it... **Vamp:** 8 Bars (*La-La-La-La-la*)

Chorus: 8 Bars ③

My milkshake brings all the boys to the yard: (4 Bars)

My milkshake brings all the boys to the yard: (4 Bars)

Verse : 8 Bars ④

I can see you're it... **Vamp:** 8 Bars (*La-La-La-La-la*)

Chorus: 8 Bars ⑤

My milkshake brings all the boys to the yard: (4 Bars)

My milkshake brings all the boys to the yard: (4 Bars)

Middle 8: (Bridge) 8 Bars ⑥

Oh, once you get involved....

Break: 8 Bars ⑦

(*La-La-La-La-la*)

Chorus: 8 Bars ⑧

My milkshake brings all the boys to the yard: (4 Bars)

My milkshake brings all the boys to the yard: (4 Bars)

Outro: Fades away ⑨

What you are almost always guaranteed, however, is an intro, verse, chorus, two breakdowns with builds and two choruses in between, and an outro. These are important, because they tell you where to start your mixes between tunes, and thus stay fairly consistent from tune to tune. But what happens between them can be anyone's guess. Thus, the number one rule of DVJing:

Know your tunes!

The language of video

I'll get deep into this when I talk about making your own DVD video single, but for now, just think about the language of what you see.

Treat it like words, sentences and paragraphs. A word is something simple, like a shot, or an edit from one shot to another. Obviously a shot can say something, but it's the edit that gives it meaning. Two people kissing says "two people kissing". But fade that to a

shot of the moon and stars, and it says "Lurve!"

Sometimes shots just say what they say, and need the edit to put them together to make sense. Take three shots – a motorcycle gang; moms in a station wagon with a baby; and an abandoned nuclear power plant. By themselves, they don't say much, but put them together into a visual sentence, and they say "here comes trouble".

Put several of these sentences together, and you get a "paragraph" of video. For music videos, I can say each section of the tune – intro, verse, chorus, etc. – is a paragraph.

Watch a music video to see how it changes from paragraph to paragraph, and how it is built around the structural sections of a song. Once you've done that, you'll realize how much you know already about making and mixing music videos.

6. MIXER AND DECKS

ON THE DECKS

CDJs and DVJs

EXCEPT FOR A COUPLE OF IMPORTANT POINTS –
ONE PLAYS CDS, THE OTHER PLAYS DVDS AND CDS –
DVJS AND CDJS WORK MUCH THE SAME WAY.

At the center of each is a platter surrounded by knobs, buttons, sliders and switches. Put your finger on the platter and you move the tune back and forth, just like vinyl. Run your finger around the platter's edge and you momentarily speed up or slow down the tune. The slider does the same thing, but the adjustment is more permanent.

Most of the buttons are there to let you pause the tune at different points, or make the tune play in a loop. The reverse switch is for backwards or forwards. The knobs change how quickly the tune starts or stops when you press one of the buttons. Here's a surprise: none of these have a real physical effect on the disc you're playing.

When you put a CD or DVD into the machine, most of the track is sucked up into RAM. The platter, knobs, buttons, slider and switch are just interface tools, like a keyboard and a mouse, to manipulate the sound that's loaded into memory.

Get it? Your CDJ and DVJ are computers tweaked to do a single thing: play music and video. When you understand that, you'll understand them as instruments, not just decks, and your days of playing a record straight through from beginning to end will be over (see **DVJ Zen Master**).

The DJ mixer

There are two ways to look at a DJ mixer: it's either the place where you fade from one tune to another, or it's a musical instrument. Either way, it's your best friend. Non-DJs think it's all about the decks. That's why you're in the booth and they're propping up the bar scratching their chins. I know the mixer is the place where it all comes together.

SIGNAL FLOW

In the analog days, big recording studio mixers were designed to take sound from dozens of mics and channel it onto eight or 16 tracks on a tape. Those eight tracks would also need to be mixed onto two tracks for a stereo recording. A mixer that could take 24 sound sources and mix them onto eight channels, and then down to a stereo left and right was called a 24 x 8 x 2 mixer. A mixer combining 16 sources on four channels, and then stereo, would be 16 x 4 x 2.

DJ mixers do the same thing. An industry standard mixer like the Pioneer DJM-800 can take sound from ten sources, route it to five channels (counting the mic channel) and mix it down to digital stereo (10 x 5 x 2) without breaking an analog sweat. Understanding this routing is the essence of understanding your mixer. It's called "signal flow".

The first thing a great DJ does when he walks up to a new mixer is try to understand the "signal flow". The same goes for a recording engineer in a new studio. It's like tuning your guitar, adjusting the turntable tonearm or blowing the spit out of your sax. Once you've done it, you're ready to play.

If you understand where you can send sound within the mixer, you can radically shape a tune. Once in a channel, sound can be sent through equalizers to shape the bass and treble, effects to create echo or reverb, and gain pots (potentiometers) to change the dynamics of the sound. You can also route mixes of several channels into your headphones, separate from the main speakers.

Every piece of software and equipment in this book has a signal flow, even visual tools. Learn the signal flow of your mixer and you'll have mastered one of the most important technical concepts of DVJing.

The video mixer

So far, none of the big boys have made a proper DVJ mixer. There's no shortage on the audio side, with every variety of DJ mixer you could imagine selling cheap down at WalMart or Woolies. For video, Edirol has put out several mixers designed for the club, most notably the V4. And Korg have released the KrossFOUR to go with their Entrancer audio/video effects unit.

But both the V4 and KrossFOUR are mute mixers, without audio. There are old skool audio/video mixers from Panasonic, Numark and a load of others, but even though some of them are "made for DJs", they often lack EQ and proper headphone monitoring. Read: they're no good for the club.

Right now, your best bet is to use separate DJ and video mixers for your DVJ nights (although this is certain to change the moment this book goes to print).

Video mixers work a bit like audio mixers, but with a few key differences. First, affordable machines rarely have more than four inputs. More important, you can only mix two of these inputs together at a time. And the output of a video mixer is a single channel, rather than two stereo channels. Most video mixers are 4 x 2 x 1. Video boffins call the two channels in the middle "busses", and they're usually referred to as Bus A and Bus B. I think channels is fine.

Why can't you mix more than two channels at a time on most mixers? I'm sure there are technical and economic reasons, but mostly it's because your video wouldn't look good. Layering one video over another is chaotic enough. Add another layer without some careful filtering, and it'll look less like *Crouching Tiger, Hidden Dragon* and more like Walt Disney blew.

Before you forget ...

... none of this matters much. It's important to learn how things are put together technically and to hone your skills. But at the end of the day you'll only be judged on one thing:

Can you rock the party?

7. CUEING, OR HOW TO START AND STOP A TUNE

CUEING DOES A LOT OF THINGS. IT LETS YOU PREVIEW TUNES BEFORE YOU PLAY THEM ON THE DANCEFLOOR. IT LETS YOU TEST MIXES. IT LETS YOU BEATMATCH. AND IT HELPS YOU REHEARSE MIXES BEFORE THE CRUCIAL MOMENT WHEN YOU START MIXING THEM TOGETHER LIVE FOR THE CROWD OUT FRONT. AND IT ALL HAPPENS IN YOUR HEADPHONES.

The basic headphone cue

While the dancefloor hears one tune, you can listen to another in your headphones. Different mixers ask you to punch different buttons to make this happen, but the result is more or less the same.

If you're beatmatching, you slide one of the cans off your ear and leave the other one on. With one ear, you're listening to the sound in the club. With the other, you listen to the tune you're cueing.

This is your chance to check a few things out:

- Are the sound levels more or less the same for both tunes? Check the level meters on the individual channels of the mixer to get them right.

- Are the high and low end balanced between the two, and do they clash? I'll cover this more in the beatmatching section.

- Are the two records beatmatched?

This is also your chance to find the place where you want to start the tune, and get it in cue (or pause), ready to play.

Setting your start point

Assuming you're starting your tune at the beginning, here's the process:

1. Slide the CD or DVD single into the deck

2. Make sure the deck is in cue on the mixer – you'll

hear it in the headphones, and the dancefloor won't

3. Press PLAY and let the tune roll until you hear the beginning

4. Put your finger on the platter and move the tune back to the beginning

5. Scratch back and forth across the beginning until you find the absolute point where the sound starts

6. Press PAUSE then press CUE.

Your tune is ready to play, and will start instantly when you press PLAY again.

When the beginning isn't the beginning

DVJs of the mobile and radio variety start tunes at the beginning. They're typically playing special "radio edits" of a tune, and they only last two or three minutes.

Beatmatching

Club DVJs almost never start a tune at the beginning. Their concern is for maintaining the flow of the music across the night. Tunes have their own arc, which sometimes disrupts what the DVJ is trying to do, so club DVJs will find places to start tunes that suit their own needs.

THE HIP-HOP AND R'N'B MODEL

There are dozens of ways to DVJ every type of music, but Hip-Hop and R'n'B nights really go off when you don't fuff about too much. Two of the UK's greatest DJs, Tim Westwood for Hip-Hop and Trevor Nelson for R'n'B, have honed this rapid-fire style of DJing to a fine art.

Intros and outros don't exist here. Tunes are cued to the first chorus or hook. Drop a tune in at the chorus and the place goes off. Let a chorus, a verse, a chorus, the middle bits and a couple of choruses play through, and move on to the next tune, dropping it in at the chorus. No slow blends, just slam the tunes back to back – and never leave them on the decks too long. Westwood goes through 30 tunes an hour – about twice as many as your average House DJ.

This isn't the kit I use at home!

When you step up to a new mixer, or any new kit in the club, don't be embarrassed to ask which buttons to push. I've seen superstar DJs grab someone (anyone!) and ask where the cue buttons are on industry standard mixers. No embarrassment there. Better to look stupid in front of the sound guy than to blow your set because you were too chicken to ask what you needed to know.

WHAT WOULD ROGER DO?

Roger Sanchez is also a master of this type of mixing, but on a House tip, which means combining tight programming with skillful blending and beatmatching, and allowing more of the tune to play.

Blending two tunes so the meat of the second tune starts where the meat of the first leaves off is an exercise in nerves and anticipation. If you start the second tune 64 beats ahead of where you want to fade out the first tune, then you have to know (64 beats from the end of the first tune) that it's time to start the second.

Slam DJs – that is DJs who slam one tune into another without blending – don't have to worry about this. They can just wait for where they want one tune to end, and start the second.

DJs who blend have to KNOW THEIR TUNES. That's really all it takes.

Rapid-fire X-fader action

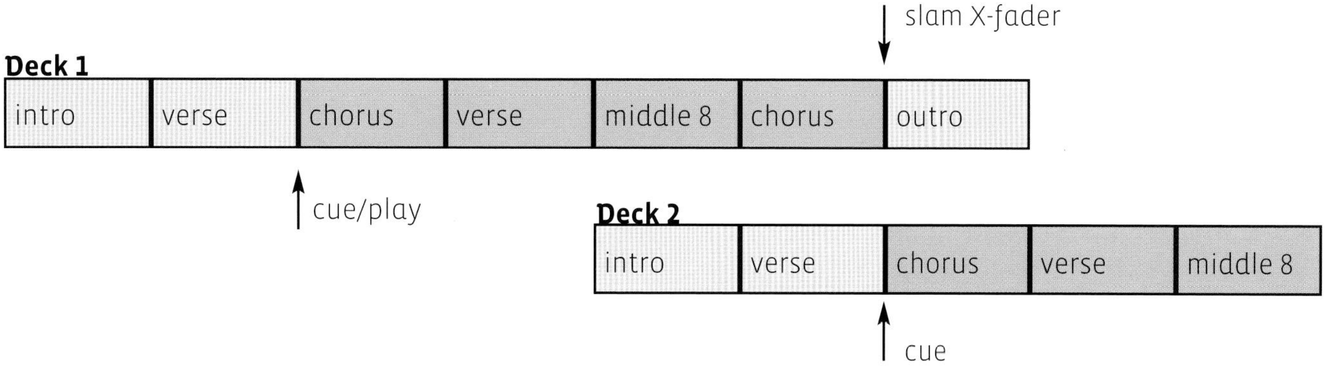

slam X-fader

Deck 1

| intro | verse | chorus | verse | middle 8 | chorus | outro |

↑ cue/play

Deck 2

| intro | verse | chorus | verse | middle 8 |

↑ cue

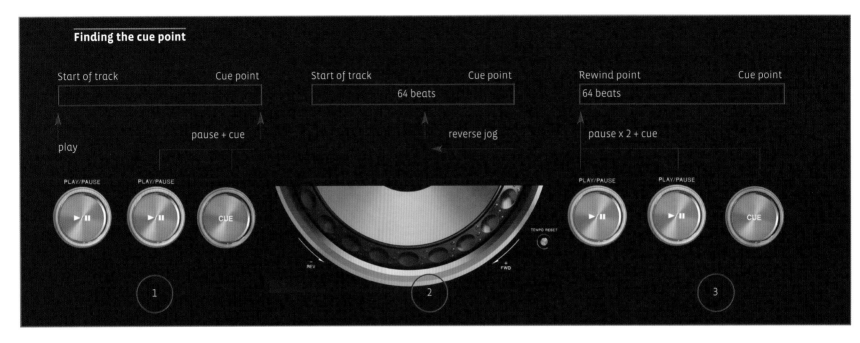

Finding the cue point

Start of track	Cue point

play

pause + cue

Start of track	Cue point
	64 beats

reverse jog

Rewind point	Cue point
64 beats	

pause x 2 + cue

PLAY/PAUSE PLAY/PAUSE CUE

PLAY/PAUSE PLAY/PAUSE CUE

1   2   3

Know your tunes, and you'll anticipate start/stop points instinctively.

64, 63, 62, 61, ...

There's still the small issue of how to cue a tune 64 (or however many) beats backwards from the point where you want to bring it in. How do you know which point is 64 beats before your starting point if it's not at the start of the track?

Here's the method: start your tune. Find the place you want to bring it in. Now follow the cueing process, except you'll be scratching back and forth across *your* beginning of the tune, not the producer's.

Press PAUSE, then CUE.

Now, put your finger on the platter, and rewind the tune 64 beats, counting them in your head. Do it slowly, counting each beat. It's confusing at first, but you'll get faster with practice.

Once you've found the rewind point, press PAUSE twice, then press CUE.

Play it back and count once to make sure you've got it right. If you have, you're ready to roll. If not, try it again.

Cue points and memory

Some CDJs and DVJs let you save cue points on memory cards. Surprisingly few new DVJs are even aware of the memory cards, but nearly all the best digital jocks use them religiously.

Saving cue points saves you time. By setting up cue points and loops in your studio for your tunes, you don't have to set them up each time you drop that tune on the dancefloor. This gives you more time to do the thing a DVJ is there to do ....

Rock the party

There's a fallacy left over from the DJ and VJ days – that jocks are in the manipulating vinyl and video business. Wrong. DVJs are in the entertainment business. They might not retire to Vegas to scratch "My Way" for grannies on gambling holidays, but it's still their job to do one thing – rock the party.

VJs are particularly bad about trying to get everyone to stare at the screen, just as some Hip-Hop DJs spend so much time displaying their skillz, they never manage to drop a beat.

I say: let's party!

Scratch, effect, mix, blend – do what you wanna do –

but drop that beat and give the audience some space to get their head down and dance.

If a piece of technology gives you more space to gauge and engage the crowd, take it and own it.

There is no cheating when you're rocking the party. It's a new dawn and a new day for digital jocks, and you shouldn't let yourself be held back by a vinyl ideology.

Rock the party

8. BEATMATCHING

YOUR KEY SKILL AS A DVJ IS TO LIFT A CROWD AND MAKE THEM SHAKE THEIR ASSES. AFTER THAT, BEATMATCHING AND MIXING RATE A PRETTY HIGH SECOND.

Listening to two sources

Beatmatching is the ability to listen to two records at once, listen to them go out of sync, decide which is moving faster than the other, and adjust them to play at the same speed without losing your marbles. It sounds impossible, but it couldn't be easier.

Your two main tools for this are your headphones and the DJ monitors in the booth.

DJ MONITORS

There should be at least one speaker in every DJ booth called the DJ monitor. It's there because you really shouldn't try to beatmatch using what you hear from the speakers on the dancefloor.

You'll remember from elementary school that sound travels slowly. If you're trying to beatmatch by listening to a speaker that's even four meters away, the few milliseconds it takes for the sound to travel the length of two basketball players is enough to make your mix sound like a pair of sneakers in the dryer.

Your DJ monitor should be as close to you as possible, and you should be able to control its volume from the mixer, separately from the volume on the dancefloor.

Your DJ monitor is where you'll get one half of the information you need to do a successful mix.

Your headphones are where you'll hear the other half.

CUEING – OLD SKOOL AND NU SKOOL

Put your headphones on so that one ear is covered

The cant of a DVJ's headphones

Some DVJs wear headphones across the top of their head, like their dads smoking a pipe listening to Liszt. Boring!

Sander Kleinenberg rides the headband low on his forehead. Schweetcore!

Fabio's style? One can on the ear, one on the forehead.

Trés cool! But try it with hair and they go flying.

Some geezers turn them upside down, with the headband looking more like the faceguard on an American football helmet. Try this with too small a head, though, and they'll fall straight to the floor.

Define your style! Check it in the mirror!

 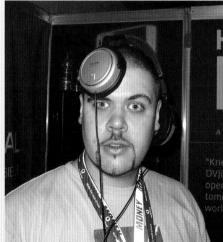 

and the other is open. Listen to the tune in cue on your headphones. Listen to the master mix on the monitors. Now try to get them set so the kick drum on the count of one happens at the same time on both tunez. One tune will drift out of sync, eventually. Try to figure out which one. Is it going too fast or too slow? Now adjust the tempo slider slightly on one of the decks and start again. Soon, you'll have them perfectly matched, and soon you'll be able to do it quickly. This is old skool DJ technique, done proper.

Some nu skool DVJs prefer to "split cue" in the headphones. This sends what's in cue to one ear and what's playing on the dancefloor to the other. The main advantage? Most monitors cause distortion, and have to be turned up louder to let you hear what you need to hear. Distortion exhausts the ear and deafens DVJs.

Another technique is to mix the master and cue sound in the headphones, sending it to both ears. I prefer

this technique, as you can actually get a fair preview of what your live mix is going to sound like on the dancefloor. You might think your Metallica track goes well with Benny Benassi, but the headphones will say "Wrong!" Better them than the dancefloor.

With the second method, you can turn the cue mix knob left and hear what's in cue, right and hear what's on the dancefloor, or put it anywhere in the middle and hear any combination. You can preview what it will sound like as the tune in cue comes in louder and louder.

Some old skool DJs dismiss this method, swearing by monitors, claiming they want to hear what the dancefloor is hearing. Here's some news: you can't hear in a DJ monitor what the dancefloor is hearing. Most monitors suck. Better off using the headphones.

Bpms

DVJs live and die by bpms. Beats per minute. Genres are classified by bpm. There are rules about bpms (never go over 128 bpm before midnight). Old vinyl DJs spent lifetimes learning to measure bpms against their heartbeat. Don't bother. DVJs just look at the counter, and spend the time they saved finding the next killer tune.

MATCHING BPMS

You have a tune playing. A quick glance at the LED panel on your CDJ, DVJ or many mixers, and you'll instantly know the bpm of that tune. On your other deck, after you've loaded your tune and started it in cue, adjust the bpm slider to match that tempo. If the bpm of the new tune isn't showing in the LEDs, let the tune play for a few bars so the deck can figure out the tempo.

Turn it down! I said turn it down!

Ever wonder why older DJs just smile and nod a lot when you talk to them. Why they don't say much back? Why when they do say something, it has nothing to do with what you said? Do you think it's because they're either a) drunk; or b) arrogant? You're wrong on both counts.

The answer is c) they're deaf!

DVJs who don't turn down the monitors and the headphones between tunes go deaf. Period.

Turn it down.

Smart DVJs use earplugs inside their headphones. That may sound counter-intuitive, but sophisticated custom-made earplugs filter out the frequencies that make you deaf, and still let you hear what's important. They cost a mint – around $150 – but they save your hearing, and your career. Watch porn with the sound off and you'll get some idea of what your future afterparties could be like without them.

Now cue that second tune. Once you've pressed start, you can instantly put your tune back in pause at the cue point by pressing the cue button again.

FINE-TUNING

In cue, start the second deck in time with the first tune. If you didn't start the deck perfectly, run your finger around the outer ring of the second deck's

BPM s and their associated genres	
BPM	**Genre**
75–90	Ambient, Chillout
100–120	Big Beat
125–145	Breakbeat
128–145	House, UK Garage, Trance
140–150	Hardhouse
145–160	Hi-NRG
170–190	Drum'n'Bass
170–200	Hardcore
180–200	Happy Hardcore
200+	Gabba, Psytrance

platter until the two tunes are in sync. Do this in the wrong direction and the tunes go further out of sync. Keep doing it until they are at least momentarily together.

Now listen. Chances are, they'll slowly creep out of sync again. Decide whether the tune in cue is going too fast or too slow. Adjust the slider. Put the tune back in cue again. Repeat.

This is really difficult the first time. It's as easy as breathing the 1000th time.

Practice, practice, practice.

THE DISAPPEARING BASSLINE

Sometimes when your tunes are perfectly in sync, you'll get the feeling you're not hearing one of them. The instinct is to make an adjustment to one of them. Don't do it, unless you want to start a train wreck.

If you feel like all you're hearing is one tune, then

you've beatmatched so perfectly, it's time to move on to mixing. If it ain't broke, don't fix it.

PERFECTION VS. ANTICIPATION

Once you've got both tunes running really close to the same speed, you'll know whether the tune in cue has a tendency to run too fast or too slow. Now you have a decision to make: should I spend another minute trying to get them perfectly in sync, or should I leave it, knowing what kind of adjustment I'll have to make when they're both playing out on the dancefloor?

That's up to you. And you'll only figure out which you prefer by rehearsing.
Practice, practice, practice.

The cheat for DVJs

Most DVJs who do a lot of beatmatching are playing tunes that don't naturally come with their own music videos, so they have to make their own videos in the studio. When they do, they have a huge opportunity to cheat. How? By setting all their tunes to the same bpm!

And so, grasshopper, I must ask the question. Is that: a) cheating; or b) giving themselves more time during the mix to manipulate tunes and drive the dancefloor loony.

And if it is cheating, is it bad? You decide, but here's a clue:

Remember the nerdy guy with the glasses who sat at the front of the class and took his tests with his hand covering his pencil so you couldn't see his answers? Chances are he won't be dropping dancefloor bombs in Ibiza this weekend.

ON THE DECKS

YOUR TRACKS ARE BEATMATCHED, AND THE FIRST ONE'S ABOUT TO RUN OUT. TIME TO START MIXING.

The mix is the smooth transition from one tune to another. Trance is the great land of the long, smooth, perfectly-blended mix. You can examine a Sasha set with a magnifying glass and still not know where one track ends and the next begins. Other DJs get creative with their mixes from one track to another, playing with the contrast between tunes, rather than going for the epic, all-night single soundtrack feel. Either way, there are a few basic skills to learn in the fade from deck to deck.

Kick it off

This is the place to be confident. Once you reach the point where you think you should start your next tune, start it. Don't hesitate or you'll miss your chance.

Now that you've started that second track, make some adjustments while the incoming tune is still in cue, to be certain your two tracks are in sync.

Upfader vs. crossfader

Which fader should you now use to bring in your new track? Should you try the crossfader, which automatically brings down one track and brings up the other? Or should you leave your crossfader in the middle and use the upfaders to manually bring one in and the other out? Which you choose depends on your style. Scratch DJs make heavy use of the crossfader. House and Trance DJs often prefer the upfaders.

If you're on the crossfader tip, just bring the fader slowly from one side to the other, listening all the while.

On the upfaders, bring one track up, then bring the other down, listening and adjusting as you go.

Keeping in sync in the mix

While you're bringing one track in and the other out, pay attention to insure the tracks aren't drifting out of sync. If they start to, make an adjustment to the outer ring of the platter, preferably of the track coming in (your punters are still focused on the track going out).

How do you tell which tune is which when they start drifting?

1. Know your tunes. This is the best way.

2. Listen to what's in cue vs. what's in the DJ monitor. If your mixer let's you blend cue and master in the headphones, go back and forth between the two, while you're in the mix.

3. Try taking your new track out of cue and putting your outgoing track into cue.

If it all starts going horribly wrong, and believe me it will at least once per gig, then grab the platter on the outgoing tune, spin it backwards and fade it down quick. This is called backspinning, and it's not only a

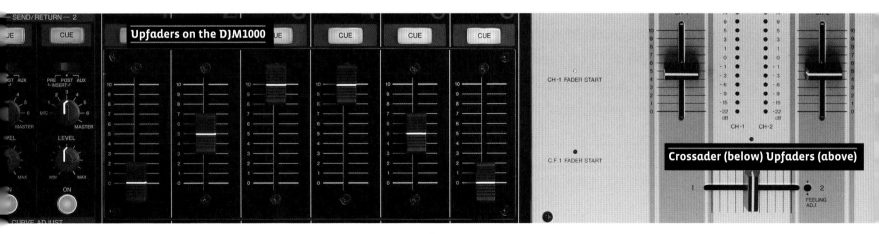

Upfaders on the DJM1000

Crossader (below) Upfaders (above)

lifesaver, it's a great piece of DVJ flair. It sounds great! Just don't do this every tune, or the punters will know you're faking it.

Varying your mixes

Sometimes two tunez should lie side by side across long phrases.

Sometimes you wanna get one in and one out quick.

Sometimes the outgoing record sticks around too long.

Listen to your tracks and decide. What's not important is that you are a DVJ who can technically mix to perfection. What matters is that you're a DVJ who can make two tunes sound like lovers.

Confidence

Eddie Halliwell is the master of this. Don't be precious with your mixes. No one likes a DVJ who is

so careful, they take away all the passion.

Eddie Halliwell

10. MIXING STRUCTURALLY

ON THE DECKS

NOW THAT YOU KNOW HOW TO BEATMATCH, AND GET FROM ONE DECK TO THE NEXT, IT'S TIME TO LEARN WHEN TO TAKE THAT STEP. IF THE BEGINNING OF THE TUNE ISN'T THE BEGINNING, AND THE END ISN'T THE END, THEN WHERE DO YOU MIX FROM ONE TO ANOTHER?

Structural mixing
KEEP THE PARTY ROLLING

Tunes are structured around what the dancefloor expects to happen next. The punters look like they're just dancing, but unconsciously they're all counting to 64, just like you.

If you want to keep the vibe alive, it's important to keep your mixes moving with the structure of the tunes. If you cut from one tune to the other three bars into the verse, rather than at the end of a section, chances are you'll kill the vibe.

PLACING TUNES

The key is to start your next tune when a section of your outgoing tune ends. That could mean starting the chorus of tune two when the final chorus of tune one plays out, often a good idea. But it could also mean starting the breakdown of tune two when the breakdown of tune one ends – probably not the best idea on the dancefloor. You'll leave your dancers in such a drifty place, they may drift right out the door.

There are no hard and fast rules about where to place a mix, but there are a few approaches that work for most dance tunes.

THE INTRO AND THE LAST CHORUS – A COUPLE OF APPROACHES
Starting the intro after the last breakdown

For most dance tunes, and House tunes in particular, starting tune two from the beginning at the same time the second breakdown of tune one ends is a safe play. Chances are tune one's choruses will play

*[The punters look like they're just dancing,
but unconsciously they're all counting to 64,
just like you.]*

out just as tune two's verses are starting. That's also the best place to mix the two, after letting tune two ride in cue through the intro until its verse starts to come in.

This makes for a loose mix, with lots of intro and outro playing out in the club, but if you're rolling a dancefloor along, it's a workable approach.

Starting the intro at the last breakdown

If you're trying to build dancefloor energy, another approach is to start tune two from the beginning just as the second breakdown of tune one is starting. In this case, you'll get a lot less intro and outro in the

---

Mixing *at* the last breakdown

Deck 1

breakdown	chorus	outro

Deck 2

intro	verse	chorus

mix. However, mixing this way could be too tight, with tune two walking all over the chorus of tune one. When this method works, it's brilliant.

Remember, these are guidelines, and every tune is different.

Know your tunes!

CUTTING BREAKBEAT

Breakbeat, although more of a dance genre than a vocal genre, is still considerably harder to mix than House or Trance. Often, the intros and outros have active high ends and fat bottoms, even before the verse kicks in. Because of this, it's not unusual to cut between tunes – or slam them one into another – rather than do a smooth mix.

Cutting here means cutting off tune one and cutting in tune two with the shortest mix possible. Cutting is a great way to build excitement, and to get between hard to mix tunes. It can also be jarring for the dancefloor – but not nearly as jarring as a car crash of high hats and fat bottoms. Use the same structural principles you would use for House and Trance.

Mixing *after* the last breakdown

Deck 1	breakdown	chorus	outro

Deck 2	intro	verse	chorus

Cutting is easiest using the crossfader, which, combined with scratching, is another reason it is the preferred mixing method of Breakbeat, R'n'B and Hip-Hop jocks.

BASSLINES

Dance music is all about big fat bass, but when you're in the mix between two tracks, turning both basslines up to full volume can be a muddy mess. You can get around this, and get more creative, by EQing the bass down on one or the other of the tunes. Here are a few methods, with variations, that work:

- Turn the bass down on the incoming tune; then

- At a good structural point, simultaneously turn the bass down on tune one, and up on tune two; or

- Two beats before that good structural point, turn the bass down on tune one, and then up on tune

two after the two beats pass; or

- Sixteen beats before your good structural point, turn the bass down on tune one, then slowly fade up the bass on tune two across the 16 beats.

This is another know-your-tunes and practice-practice-practice moment. And don't get too heavy-handed on the bass knob. Bringing the bass EQ fully down also takes out a lot of the lower mid-range, and can sound too harsh. Listen and do what sounds good to you.

EQ IN THE MIX

What's true for bass can be true for mid-range and treble. Tunes with heavy instrumentation in the mid-range clash against one another in the mix, and mid-range EQing will save your ass. Same goes for the top end. Listen and do what sounds good to you.

EQing the bass – example 1

start of 4-bar phrase   good structural point

Channel 1
Bass Eq

Bass

Beats

Channel 2
Bass Eq

Bass

EQing the bass – example 2

start of 4-bar phrase   good structural point

Channel 1
Bass Eq

Bass

Beats

Channel 2
Bass Eq

Bass

EQing the bass – example 3

start of 4-bar phrase   good structural point

Channel 1
Bass Eq

Bass

Beats

Channel 2
Bass Eq

Bass

EQ FLAIR

While you're thinking about EQ, it's good to remember that EQ is another in a range of audio effects, and therefore is one of the DJ's main expressive tools – like cymbals to a drummer. Think of EQ in this way, adding a little extra oomph to the bass on the one, or high end twiddle in the cymbals on the upbeat of the four leading into a new section of your tune.

Practice and see what works for you. Don't forget to turn your EQ knobs using your whole body, arm in the air, elbows pointing at the ceiling then pointing at your shoes. The punters will know you're doing something.

PUTTING THE 1 ON THE 1

When you start a new tune, it's best to start at the beginning of a section of the last tune. That also means starting it on the count of one. Start it on the two and you'll confuse the dancefloor about where the one is. Remember, they're counting **1**, 2, 3, **4**, 1, 2, 3, 4.

STACKING THE 3 ON THE 1

You can try starting on the three instead. Then you'll

Stacking the 1 on the 1

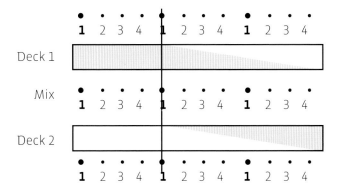

Stacking the 1 on the 3

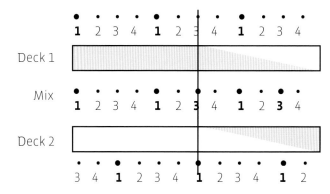

get a rhythm that rolls along as **1**, 2, **3**, 4, **1**, 2, **3**, 4. I've heard great DJs do this, but it doesn't come without a lot of practice, and is usually best left for laying one tune over the top of another for a single section, rather than mixing from one record to another. Practice and see.

Mistakes

The most important DVJ lesson I know, I learned from a lounge singer I was sharing a drink with after a gig in Kuala Lumpur. He had been playing Asian lounge bars in a tuxedo for decades, and had a voice like Jim Beam and Marlboro Reds. He said this:

"If I make a mistake, when the repeat comes back around, I sing it wrong again. Then no one knows I mucked up the first time."

No one knows what you're doing better than you. Think about that.

They're counting 1,2,3,4

1,2,3,4

1,2,3,4

1,2,3,4

1,2,3,4

my hero

1,2,3,4

ON THE DECKS

WHILE YOU'RE IN THE DJ MIX, YOU'LL NEED TO MIX AND CUT THE VIDZ TOO. THE CONCEPT IS SIMPLE – SWITCHING FROM ONE VIDEO TO ANOTHER, THE SAME AS YOU WOULD WITH AUDIO – BUT THERE ARE A FEW KEY DIFFERENCES.

T-Bar vs. crossfader (V4 vs. KrossFOUR)

Video gear for DVJs is still largely based on video gear for television producers. That's why many video mixers have a big T-shaped fader where you might expect a DJ-style crossfader. It's called a T-bar. On a big BBC video console, it's heavy with a weighted pull for slow, smooth transitions.

The Edirol V4 has a smaller, plastic T-bar that still shows some resistance when you pull from one video to another. The friendly engineers at Edirol have also added the option of turning the T-bar 90°, so it seems to function more like a crossfader. The T-bar is great for transitions, but not so hot for scratching, and some VJs have replaced their T-bars with crossfaders

to make video transitions. Do this at your own risk, and make sure you know at least a little something about electronics before you give it a go.

The Korg KrossFOUR already goes the crossfader route, although there is still some resistance in the transition. Which style fader you use is really a matter of preference and budget. Try them both before buying your gear.

Cuts

The most straightforward transition from one video to the next is the cut. Because of the way video mixers are designed, all your sources are available on every channel and are selected by punching a button. Cuts are done by pressing the input source you prefer. Some high-end video mixers do this smoothly, but lesser mixers – the V4 included – will make the picture jump when you switch inputs. To make up for this, the V4 has a transformer button that will cut from one channel to another without losing sync (making the picture jump). The ultra-affordable VSW-1 also cuts smoothly.

Fades

A fade from one video to another is done on the T-bar or the crossfader. It's a slow transition, just like a crossfade or mix on the DJ mixer. Many mixers will also automatically (and very smoothly) fade at the touch of a button.

Wipes

Wipes are ugly and are not your friend. A wipe brings one video up and the other down with a hard-edge in between, kind of like a windshield wiper pushing one piece of video out and bringing another in behind it. Think of old episodes of the original Batman TV series and you'll know what I mean.

I'm sure they can be used effectively and creatively, I've just never seen anyone do it. Ever.

VJ mixers

Here I have a look at the Edirol V4. VJ mixers share many features with DJ mixers – they're specifically designed to be club friendly. Here I have a look at preview functions, fading from video to video, selecting your input source, adding effects, and the bus system.

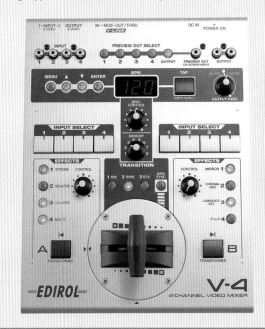

Layering with fade

It's possible to have two videos playing at the same time, one on top of the other, just as you can have two audio tracks playing simultaneously to create a third. You can do this with the fader by putting it in the middle, as you would put the crossfader in the middle position on your DJ mixer.

This can look nice and is often none-too-jarring, depending on the content of the videos, but it comes at a cost. Neither video will play at full brightness, so your colors will be a bit muddy. Practice and decide which videos work well this way. Many won't.

Layering with chromakey

Chromakey allows you to mix one video over the other based on areas of color. Chromakey removes a range of colors from one video to reveal the other video behind it. The removed colors become transparent.

This is how they put the weather girl in front of a weather map.

The fade out

Chromakey is a great effect because both videos will come through at full brightness. Unfortunately, to make chromakey work effectively, you need to shoot the video you want to put on the top against a blue

Photosensitive epilepsy

Photosenstive epilepsy is a rare form of epilepsy that can induce seizures in sufferers when they are exposed to rhythmically flashing lights, videos or patterns. It's common in younger people (read: your audience) and can manifest itself in people who've never experienced an epileptic seizure. Strobes, fast-cut video or even watching a train as it passes in a station can induce deadly seizures.

In 1997, hundreds of Japanese Pokémon fans fell into simultaneous seizures across the country when their television screens flashed for several seconds with video from the cartoon. Every major broadcasting organization in the world subsequently put in place a policy concerning the broadcast of flashing images.

At the BBC, no more than three significant changes in luminance (brightness) levels are allowed per second. This is a rule you should follow. Ending a fan's life is no way to end a set.

or green screen with very even lighting. If you don't have access to a big television studio, you probably won't be doing this anytime soon.

Layering with lumakey

Lumakeying is the better option for DVJs. Lumakey lets you bring one video over the other based on areas of brightness. This way, you can make the dark areas of the top video disappear to reveal the video beneath. Lumakey is one of the main effects for creative DVJs. It's also a great way to layer a logo with a black background over another video.

Avoiding video chaos

If two clashing tunes played one over the other create a train wreck, two clashing videos create nothing more interesting than mud. In some cases, particularly when using keying rather than fades, both videos will appear to go abstract, and the wash of images confuses the eye. This can be a great

Chromakey

effect, or it can be mind-numbing if used for too long. You can only decide in the rehearsal studio. As with tunes, know your videos. And practice, practice, practice!

Quick cutting

Quick cutting from one video to the next in a scratch style is not really effective when you're using either a T-bar or a crossfader with resistance. And sometimes trying to scratch a track, handle effects, get the

sound right and grab the T-bar can make you wish you were born an octopus. This is one of the main arguments for using as many technological cheats as possible. The V4 has a transition effect which will cut or fade back and forth between videos based on bpms. This is effective for scratch DVJs who want to use one hand to scratch and the other to handle to crossfader or upfaders on the DJ mixer.

Lumakey

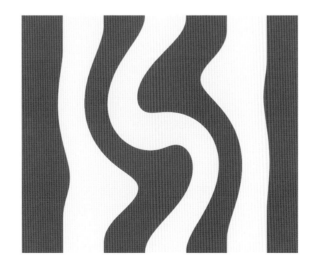

12. MIXING CDS AND DVDS

ON THE DECKS

The three-deck mix

LAYERING TRACKS

Layering one track over another is one of the key tricks for DJs. The reward for finding that special mix of a tune is all the more sweet when you've created it live for the dancefloor.

Often a track like Daft Punk's *Burnin'* can be layered over the middle and end of one track and the beginning of another, creating two new tunes in the offing.

How do you do this with two decks? You can, but it ain't as easy as on three.

DJs like Slam and Tim Sheridan of Dopesmugglaz made the three- and four-deck mix a trademark. Now that every DJ booth has two vinyl decks and two CDJs, a three-deck mix has become standard.

Still, lots of DJs won't touch it. Why? Keeping two decks in sync is a difficult enough task. Keeping three in sync can turn into chaos.

And if you haven't read the chapters on song structure, don't even go near it.

There are, however, some options beyond three decks. Mixes which also use DJ software like Ableton Live or Pioneer's *DJS* software take advantage of the automation tools available in software. Two tunes (or more) can be layered on the laptop. The software will automatically beatmatch them and keep them in sync, giving you the headspace to add a third or fourth layer from the decks, and keep everything in sync. This is a signature effect for DJs like Switch, and is well worth exploring.

LAYERING VIDEO

DVJs add a new element to the layered mix. Smooth transitions between videos can create great new video effects by layering one set of video across another. And just as the bassline from one tune goes beautifully with the top end of another, a dancing figure from one video layered across an abstract sky from another creates a third new video you never knew you had.

Some video-only DVDs in your "record box" might include smoke trails, abstract backgrounds, logos or dancers, made specifically for layering, as are many on the *How to DVJ* DVD.

Combine layered video with layered audio and every time you step up to the decks, you'll create a fresh new AV show that'll surprise even you.

Try to do it with two decks, though, and you'll give yourself a stroke.

MEMORY CARDS AND SLOW CUEING DVDS

With a combination of CDJs and memory cards, CD DJs can work infinitely faster than their old-skool

vinyl peers. Vinylistas are known for sticking tape across the grooves of certain vinyls to mark cue points. This lets them fast-switch tracks on two decks and create something entirely new.

With cue points, CD DJs are able to accomplish the same at the press of a button. Pre-marked cue points are stored on memory cards, and each time the CD DJ slots a new disc into the CDJ player, the cue points are nearly instantly accessible.

Unfortunately, with DVDs, the issue is a bit more complex. Video eats up considerably more data, so DVDs are slightly slower to load in DVJ machines. It can take up to a minute to load a new disc and call up either the cue or loop points that are also available on the DVJs. For a quick-cutting DVJ, this is just too slow.

There are, however, two simple solutions, but one costs some time and the other costs some money.

PRE-PLANNING AND PRE-PRODUCTION

This is the time-consuming but cheaper approach to solving most slow cueing problems. Think through elements of your set and do some of that fancy mixing in the studio. It's common for DVJs to have little clusters of three or four tunes that they know will work well together. They can then construct their set in the club from a variety of combinations of these clusters.

If you know what your clusters are, you can do some pre-production to make them easier to execute live:

- If you have four tunes that you know will work well together, put tunes 1 and 3 on one DVD, and 2 and 4 on another.

- Premix two tunes you like together and play that mix out in the club. Then use the extra time you have to get creative with the video on top.

These are not the ideal solutions. But ideals are often pretty damned expensive ...

THREE DECKS

The more straightforward solution is to get a third deck. This solves two problems – slow cueing DVDs, and layering both music and video.

With three decks, you've always got one deck in cue. Even if you're scratching or layering with two of your

The three-deck mix

Deck 1 – an ambient groove

video

bassline

audio

Deck 2 – track 1: dancer and vocal

video

vocal   new vocal

audio

Deck 2 – track 2: clubbers and vocal

new DVD loaded

Deck 3 – logo

| HOW TO DVJ | HOW TO DVJ | HOW TO DVJ | | | | |

video

no audio

audio

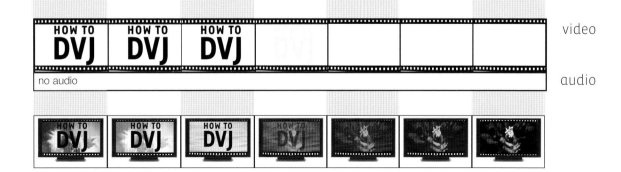

decks, the third deck can load up a new tune while you go through your routine. You might feel the need to acquire a third hand to handle all those discs, but there's nothing more pleasurable than the luxury of a third deck.

Also with the third source – even if it's a CDJ or a laptop paired with two DVJs – you'll have a more broad set of options for layering audio and visuals. In a three DVJ configuration, on decks one and two, you can layer a vocal over a chugging bassline, while on the third deck, you're layering your logo over the image of the vocalist singing on deck two. You can then drop the vocal out, fade the video to the logo exclusively, let the groove roll on deck one, and cue up a new DVD track on deck two. Fade the video from the logo on three to an ambient groove video on one, and then take the whole thing – audio and video – to deck two.

Another option is to use decks one and two for straightforward audio/video mixing from DVDs, and to use deck three for dropping in samples and scratching. I often leave a scratch DVD looping in deck three through a whole set, mixing from one to two and back, and dropping in the samples and scratches like little drum solos all night long.

A third trick I love is to mix back and forth across two decks, and scratch through the mixes from the third. This keeps all three faders up at the same time, while you're twisting and mixing on the third deck. The crowd may not be able to see it, and it sure takes a lot of practice, but you'll have the pleasure of scaring the hell out of the DJ standing behind you waiting to play next.

MIXING MEMORY CARDS
A word of warning about memory cards: the formatted ones that work in your CDJs won't work in your DVJs. You'll have to set up your cue points again when you make the transition to video.

ON THE DECKS

DJ vs. normal

On the back of your CDJ or DVJ, you'll find a switch to put your deck in either DJ mode or normal mode. In DJ mode, everything you ever thought of doing with vinyl, you can do with a CD or DVD. Everything and more.

In normal mode, your CDJ or DVJ has phenomenally high quality output, a range of connection settings, professional build quality and stunning design – but it's still just a basic CD or DVD player.

I think you'll set the switch to "DJ" and never move it again.

"Vinyl" vs. CDJ/DVJ

On top of your deck, there's another button to turn "vinyl" mode on and off. A glance at the jog wheel and you can tell "vinyl" is on, because the big neon in the center of the jog wheel display will say "vinyl".

"Vinyl" gives you all the control over your disc that you would have over a 12" on a turntable – it lets you grab the jog wheel and scratch. Why would you turn it off? With "vinyl" off, you can nudge the top of the jogwheel forward and back to momentarily speed up and slow down your tune, just like you'd nudge the label on a 12". With "vinyl" on, you can do the same, but you have to run your finger around the outside ring of the jogwheel – touching the top stops the disc.

Lots of DVJs leave "vinyl" off when they first start using the CDJ or DVJ. I think you should leave it on and learn to use it, especially if you want to take advantage of one of the key features of digital decks – loops and cues.

One other time to turn "vinyl" off is when you have a pushy guest in the DJ booth. Any slip on the jog wheel can stop the show.

Cue

When you cue a record, you drop the needle in a groove and start the deck, listening for the point where you want to start your tune. When the point arrives, you grab the vinyl with your finger, let it spin on the slipmat, and rock it back and forth across a beat until you find the right spot. Once it's found, you stop the deck, holding on to the point with your finger while the platter continues to spin. The slipmat lets the record stop while the machine keeps running.

You can do the same thing in "vinyl" mode on your digital deck. Start the disc, listen for the right point, and rock it back and forth until you've got it perfect. Then press PAUSE and CUE. That's it. You've cued your disc.

What the CDJ and DVJ now let you do that's so special is this: while in PAUSE, touch the CUE button and a cue point will be set. Pressing PAUSE will start play. Pressing CUE while the disc is playing will take you back to your cue point. This is a great way of

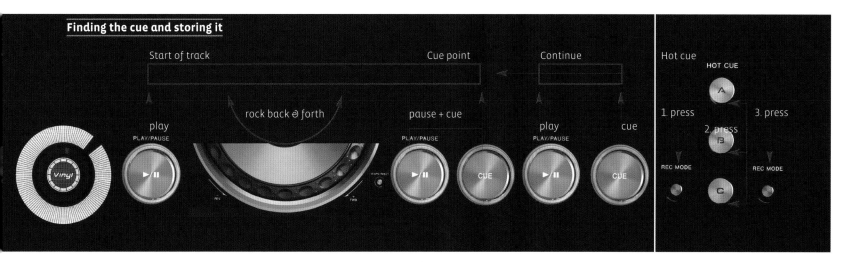

Finding the cue and storing it

Start of track — Cue point — Continue — Hot cue

rock back & forth

play — pause + cue — play — cue

1. press — 2. press — 3. press

cueing up a tune, testing the tempo and quickly getting back in cue. But it's also a great way to fire off samples from your disc!

Also when the deck is in pause and cue, by pressing the CUE button the tune will play until you let up on the button – then automatically return to cue. This is an even better way to fire samples, but takes a little getting used to. Have a play and you'll get it.

Hot Cue

The DVJ and some of the CDJs let you go a step further – you can also save and call three Hot Cue points on the "A, B, C" buttons. These will let you set up three points to return to quickly so you can fire off samples, change the song structure, stutter a vocal – almost anything you can think of doing with three samples of infinite length.

To set a Hot Cue, press the Rec Mode button, then at the appropriate point press the A, B or C button you want to store a point in. You can do this in cue as well. Or in the middle of a loop to store the loop. When you press Rec Mode again, you'll be in Call Mode and ready to call up your Hot Cue.

Loops

The DVJ and some of the CDJs also let you set up loops. Once you really start getting into loops and cues, your decks will become musical instruments.

You set up a loop by letting your track play until you want the loop to start and then pressing the "In" button. When you want the loop to end, press "End".

Chances are your loop isn't perfect, but no worries. You can adjust it by pressing either the IN or OUT button on the CDJ, or the OUT button on the DVJ. When it starts flashing, turn the outside ring on the scratch wheel and you'll hear the corresponding loop point adjusting. Once you've found the right point, press the flashing IN or OUT button again.

To get out of the loop you've set, just press Reloop/Exit. To go instantly back to your last loop from any point in the track, press Reloop/Exit again.

Before you get out of the loop, though, you might want to try saving it into a Hot Cue point if you're using the DVJ. While you're looping, press the Rec Mode button, then press the A, B or C button that you want to save your loop into. Press the Rec Mode button again. Now your loop positions are stored in

one of the A, B or C buttons, and you can use the others to store either cues or loops.

Memory cards

Setting up cue and loop points is simple, but in a set, by the time you've set up all the cue and loop points you had in mind, your other track may have run out. There you'll be with your head down, headphones on, ready to knock them dead with your on-the-fly rework of some hot track, and you'll suddenly look up to see a dancefloor full of

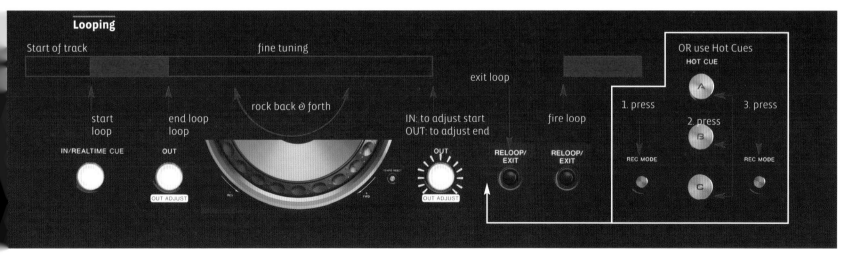

angry punters wondering why you stopped the music.

This is where memory cards come in. Both the DVJ and some CDJs let you store your cue or loop points to memory cards. When you slide a disc in, the deck will automatically find the information about in, out and cue points for that disc, and have them ready for you to call up.

On some CDJs, you can store cue points in internal memory, but I don't see the point of that. The beauty of using a memory card is that you can carry a couple of them in your wallet, and slip them into the CDJ or DVJ in the club. That way you're not carrying your deck to the club just so you can use your cue points.

Flash memory cards are the size of a postage stamp. They can store massive amounts of information about cue and loop points. For each disc, 100 cue and loop points can be stored, as well as three Hot Start

points and 20 WAVE tracks. An 8 MB card can store points for up to 5000 discs and a 16 MB card holds info for 10000.

The CDJ and DVJ have vastly different ways of accessing the data, and you'll find all this information in your manual. Just a couple of notes, though: first, once you start using memory, you'll never go back; and second, CDJ- and DVJ-formatted cards aren't compatible, so don't count on bringing in all your old CDJ cue points when you get your first DVJs.

Finally, you can also back up and manipulate your memory cards using a computer. If your PC or Mac has the right card reader, you can drag the folders from your card onto your hard drive for permanent backup, and then you can drag them onto another memory card for quick duplication. The DVJ will duplicate cards as well, but it's a rather slow process.

14. CHOOSING YOUR SET

ON THE DECKS

Myths

DVJ MYTH #1:

Great DJs and DVJs never plan their sets. They arrive at the venue with a box full of tunes, read the crowd and select their opening salvo. Once they've fired their first shot, they lead the dancefloor on an improvisational journey, the punters riding waves of excitement instinctively delivered by the DVJ god-master in the booth.

DVJ MYTH #2:

Only beginners plan their sets, and those are usually based on the playlist of some superstar DJ.

DVJ MYTH #3:

Planned sets only work in the bedroom.

DVJ MYTH #4:

Real DVJs don't pay attention to what other DVJs and DJs have in their playlist.

The real deal

Set planning is a reality, and there's nothing wrong with it – no matter what the writers at all those DJ mags say. DJs and editors like to believe sets aren't planned, and superstar DJs like to promote that myth. In the days of vinyl, when many DJs were doing little more than playing and mixing tracks, promoting the illusion that DJs were somehow akin to jazz musicians or gods was in their best interest.

My favorite quote about a superstar DJ came from one of the most powerful dance promoters in the world: "He's the only DJ I've ever seen walk into the booth and actually go into a trance." Har har!

DVJs are special, and I won't deny it, but they're also skilled performers. Performers rehearse. Rehearsal is the process of finding out what works and what doesn't. Knowing that means knowing, to a small degree at least, what you're going to do in the club.

Degrees of improvisation

There are several degrees of improvisation, from total freeform where you haven't got a clue what you're going to play when you arrive in the venue, to the absolutely set set, where you know every move you're going to make from the moment you walk in the door until you leave.

As you read over these, here's something to keep in mind: every way of working is valid!

The second worse thing a DVJ can do is get caught up in a bunch of self-imposed rules they think they have to follow if they're going to be a "real DVJ". The worst thing they can do is get caught up in someone else's rules.

TOTAL IMPROVISATION
What it is
Total improvisation for DVJs is something of a myth. No matter how free your set might be, you're still limited to the tunes you pack in your box. That aside, the totally improvised set is a free-flowing adventure. You choose your first tune, which will lead you to four or five potential choices which you feel will work well as your next tune. That choice in turn leads you to several more. Beginning-to-end is truly a journey, for both the DVJ and the audience.

One other note: a totally improvised set rarely happens in clubland.

Advantages
- There's nothing like the adrenalin rush of terror you'll feel when you step into the booth

- No two sets will ever be the same

- You'll really *really* need to know your tunes

- You'll be as thrilled as the punters when you get it right.

Disadvantages

- There's nothing like the adrenalin rush of terror you'll feel when you step into the booth

- No two sets will ever be the same

- You'll really *really* need to know your tunes

- You'll be as gutted as the punters when you get it wrong.

IMPROVISATION AROUND A THEME

What it is

Improvisation around a theme is about choosing what you're going to put in your tune box based on a number of factors, and then playing a freeform set. It can be as simple as choosing the tunes you'll take with you based on the music policy of the night, who's playing before you and after, and what time you're going to go on. If you're an international DVJ, you might base your selection on the preferences of

the dance scene in that country: do Germans like it four on the floor with lots of long breakdowns? Do the Swiss like lots of lyrics and breakdowns? Do the English like the space to get their heads down and get on with a rolling groove?

iDJ magazine once held an anniversary party where they chose the 100 classic dance tracks of all time. They then invited a cabal of DJs to play, but with the caveat that they could only play from that list. That's real improvisation around a theme.

Advantages

- Themeing is far more fun than total improv – you'll feel more confident with what you're doing

- It's always comforting to have a few rules

- It forces you to think about your audience before you arrive.

Disadvantages

- You really might not have the right tunes or videos, so a lot of preparation is important

- You really have to know your tunes

- You might not get to play the tracks you truly love.

IMPROVISATION AROUND SET PIECES

What it is

This is my preferred way of working. You have three or four tunes that you know work well together. You practice mixing that tiny set to perfection, with lots of DVJ flair and style over the top, and maybe some effects and scratching. You create dozens of clusters of these mini-sets. You arrive at the club, decide which cluster to start with, and let that cluster lead you to the next and the next.

Advantages

- You can read the crowd and still perform

rehearsed mixes

- You can stay flexible (see "The set set")

- Most accomplished sets are performed this way.

Disadvantages

- There are no disadvantages.

A SET PIECE WITH SOLOS (SCRATCHING)

What it is

Setting your set is just that – deciding in the studio what you're going to play in the club, from start to finish. Creating a set piece with solos is giving yourself the room within this to drop in some samples and scratching.

Advantages

- You'll walk into the club with absolute confidence that you can perform your set flawlessly

- You'll still have space to keep the energy high when you get the samples and scratching right

Disadvantages

- It could all go horribly wrong! (see "The set set").

THE SET SET

What it is

The set set is like walking into a club and playing a tape, but doing it live. You've decided your tunes and their order, you know exactly when you're going to drop and mix them, and you've rehearsed your scratching to a tee. Beginner DVJs do it, lacking the confidence that they have any space for improvisation.

There are many times when a set set is absolutely appropriate. Superstar DVJs, who almost always headline, often set their sets with the confidence that whatever they play, the punters are going to love it. If your set is being recorded for release as a mix DVD, or if your set is being broadcast, then setting your set is

a good idea. I've even seen a superstar DJ slot a set-length pre-mixed CD into his CDJ, press play and mime the rest on the vinyl decks. Disingenuous, true, but it worked, and he rocked a party of 15,000, plus broadcast it to the whole of the UK!

Advantages

- You won't make any mistakes

- Your set will be perfectly crafted.

Disadvantages

- It not only could go horribly wrong, chances are it will. Here are just a few ways:

  – You arrive late and an hour is cut off your set

  – The next DVJ arrives late and an hour is added to your set

  – One of your DVDs is scratched and you can't play it

– You expected three decks and there are only two

– You arrive and the warm-up DVJ has already played five of your best tunes

– From the isolation of your studio, you misread the audience: they hate tech house, and that's the basis of your set

– From the isolation of your studio, you misread the audience's reading of you: they hate you, and you're going to have to win them over. Your set set isn't going to do it.

ALL OF THE ABOVE

One last word of warning about set sets. I've played them and they've worked perfectly. I've played them and they've been disasters (all of the above disadvantages have happened to me, for real). Unless you really know the venue and the crowd, and are absolutely confident nothing will go wrong, avoid the set set.

Combining all of the above is the ideal in our minds. You can set your set, but do so around small mini-sets based on a knowledge of the night's regulars, and with the backup of a full box of tunes.

HOW TO PLAY THE LAST GIG OF YOUR CAREER (OR, HOW TO NEVER EVER GET BOOKED AGAIN EVER – FOR A WHILE ANYWAY)

Play the wrong records at the wrong time of night for the wrong crowd!

THE WORST THING THAT CAN HAPPEN TO YOU AS A DVJ

Arrive at the venue for a set that starts at midnight, and find that the warm-up DJ is flailing the audience bloody with a banging Hard Dance set. It's only 11:30 and he's already up to 150 bpm.

Believe it or not, it happened to me while I was writing this book.

It ruins your set. By the time you're on, the audience is ready for a break. You've got to bring them back down to about 135, just so you can build them back up again.

If you're a warm-up DVJ and try to play a headline set, you are guaranteed to never get booked in that venue again, or to warm up for that DVJ again, ever. You're also guaranteed that the star DVJ is going to dis you to the promoter and anyone else who'll listen when he

finishes his set. He'll feel like he didn't play his best set since you put him at a disadvantage – and he'll be right.

Eddie Halliwell, one of the UK's top star DVJs, got his start by warming up – just like everyone else. Aside from possessing mind-blowing talent, one of his secrets was that when he warmed up, he always did the job! He played the right set for the right time of night in the right club.

I don't like rules, but here's one that's golden: go over 128 bpm before midnight at your own peril.

[Roger Sanchez]
The DVJ has opened up a new world for me in my performance. When putting together my live band I wanted to incorporate visuals with the live aspect of the show, yet include the DJing aspect of the show, and the DVJs were the perfect solution. It's the future!

pic: Lisa Loco

IF AUSTRALIA'S PRE-EMINENT DJ, PHIL K, IS ABOUT ONE THING, IT'S INNOVATION. A STAPLE IN THE WORLD'S FINEST DJ BOOTHS, PHIL HAS TOURED THE PLANET WITH HIS OWN PRODUCTIONS, AS WELL AS THE CREAM OF THE REST. IN 2004, HE RELEASED A DOUBLE MIX CD WITH DAVE SEAMAN: "RENNAISSANCE PRESENTS THERAPY SESSIONS." WHILE MOST OF THE WORLD'S DJS WOULD BE HAPPY WITH THAT ALONE, PHIL WENT ON THAT YEAR TO: PRODUCE A PRESTIGIOUS MIX FOR ANNIE NIGHTINGALE'S BBC RADIO 1 SHOW; FIND HIMSELF IN THE TOP 5 RANKING OF AUSTRALIA'S DJ MEDIA AWARDS; AND FELL IN LOVE – WITH THE PIONEER DVJ-X1. IN 2005, HE KEPT THE BALL ROLLING, LAUNCHING HIS Y4K MIX FOR LONDON'S FABRIC. ONWARD AND UPWARD FOR OUR MAN DOWN UNDER!

Kriel: How has DVJing had an effect on your performances?

Phil: DVJing has helped put more of my personality and taste into my performances. Most of the clips I make myself and most of the visuals that I use are filmed by me on my travels, so in a weird kinda way it's like the audience is watching the world through my eyes and experiencing something I have while I'm playing my music.

Kriel: What is your approach to making video?

Phil: It's different every time I guess ... I listen to the track sometimes and try to think what I would like to fit with it. Sometimes I'll watch a video and think,

"Wow! That bit will really work with that track." Sometimes I'll be filming and will see something that will make me think of a track. So it's always different.

Kriel: Do you work solo or with other artists with regard to combining DJing and VJing?

Phil: I don't really work with other people, but I would love to. My graphic art skills aren't quite up to scratch and I would love to work with people who are as experienced at that as I am at DJing and producing music.

Kriel: How do you think DVJing helps you project yourself as a personality?

Phil: It allows you to project your overall taste a lot more than just playing music ... and really when you are DJing that is what you are doing. With video you can project a more well-rounded personality because, as humans, we have more than hearing as a sense.

Kriel: Do you have any tips and suggestions for people getting into this?

Phil: Don't let all the fear of technicalities get in the way of expressing yourself. I was new to this even a year ago. Even now I don't really rate myself as a video artist as my art is quite lo-fi and simplistic, but I guess that's what also makes it work and most importantly it's mine. Technically it has never been easier to express yourself with this medium, so just do it!!

IN THE CLUB

RECCE = RECONNAISSANCE, AND IT'S ONE OF THE MOST IMPORTANT PARTS OF PREPARING FOR A DVJ GIG. THERE ARE THREE KINDS OF RECCE YOU'LL WANT TO DO BEFORE YOUR GIG: THE TECHNICAL RECCE, THE GIG RECCE AND THE BUSINESS RECCE.

The technical recce

DJs seldom have to check out the club over technical issues. DVJs must.

Two decks and a mixer are easy enough to suss out over the phone. And you can't do anything about the sound system unless you're bringing your own.

Once you get into the video mixers, effectors, extra kit, screens, cabling and all the other extras that a DVJ requires, a quick visit to the club, preferably while they're closed, is one of the best ways to insure your night runs smoothly.

It's hard to make an impressive entrance for the punters once they've already seen you crawling on the floor running cables because you didn't recce ahead of time. I once found myself running around the rail-less rafters of Manumission with a projector strapped to my back and 10,000 mad punters 50 meters below me, all because I didn't get a proper recce, and the (outside) technicians couldn't sort it out.

I suggest arranging a technical rider with the venue ahead of time. Tell them exactly what you need, what you'll bring, what you expect them to provide, etc. Don't expect to get it all. Do expect to be told you'll have it all, but on the night, it fails to materialize. Nonetheless, it's worth sorting things out with the venue owners or promoters up front and in writing. They'll respect you more, you'll sweat a lot less and there's far less chance of anything going wrong on the night.

The gig recce

The gig recce is about knowing what to expect from the crowd and other DVJs on the night. All of this helps you decide how you're going to handle your gig.

The first thing you'll want to know is the music policy of the night or event. This could be anything from straight-up Top 40 to one of the myriad permutations of House. In the US, dance music is often divided into broad genres like Hip-Hop, Electronica or Trance. In the UK and a lot of Europe, you'll need a microscope to find the fine lines between dance nights, and although there are dozens of what seem like arbitrary classifications, audiences and promoters are adamant about the styles.

You'll also want to know who is playing before you and who is playing after. What time is your set? How long? Are you in the main room? Then you'll need to get people dancing. Are you in the backroom? Then you may need to just keep people's heads nodding at the bar.

The business recce

This is the contract, whether it's written or spoken. Get it in writing if possible.

Things you'll want to know:

How much do I get paid?

Your reputation and ability to put bums on seats is the principal factor. You also need to know who is paying your agent – does the promoter pay a booking fee, or do you pay a percentage, or both? Who is paying for taxis, transport, meals, hotels, extras, etc.? Under what circumstances might you not get paid, or worse, have to give the money back? What happens to the booking fee if either you or the venue cancel?

When do I get paid?

The standard is to get one half of the money on booking to secure the date, and one half of the money when you arrive at the venue or the airport if you're being picked up by the promoters. If you agree to payment after you play, you will never see the cash. Some DJs refuse to leave the airport without second half payment. Many promoters will book DVJs, and hope to pay them from the money they collect on the door. If the door take sucks, you won't get paid. You may think that might ruin the promoter's reputation, but in truth, if the door sucks, they probably won't be promoting another night at that venue. And never, ever,

ever take a cheque for the second payment. That cheque may bounce quicker than your set.

Where will my name be on the bill?

You want it as high as possible. Some nights don't advertise talent at all, preferring to promote the vibe of the night. Manumission in Ibiza is one example. This is great for the night, but not so great for the DVJ. Unless you're getting a whacking huge chunk of change, or your name-face recognition is so high you're familiar from a mile off, insist on decent billing.

Will they meet my personal rider?

Are you going to have to buy your own drinks? Is there somewhere for you to hang out while you wait to go on? Is there somewhere safe to put your kit before and after you play? Is someone going to be there to walk you through the crowd and help you carry your stuff when you arrive?

If you can get away with it, ask for as much as possible. The more you reasonably ask for, the more you'll get. The less you ask for, the less the promoter will respect you. The operative word is reasonable. No one likes a prima donna. To that end, if possible, always bring an assistant: to drive you, to fetch your drinks and to collect your money, but particularly to

play bad guy with the promoter so you have the space to play good guy. I know DJs who are universally loved and respected, but have managers who are equally loathed. That's exactly as it should be.

[Brett Belcastro – 2nd Nature]
My performances have excelled significantly since the DVJ technology was released. The DVJ technology has helped me make the clean transition from simply a turntable DJ to an artist. Timing is key for making creative cuts and transitions during a performance. DVJing gives me the ability to manipulate a party-goer's experience through sound and sight, which allows me to express much more of my own artistry.

photo: Jabu Brown

IN THE CLUB

Power

Euro, British, Australian, 240, 110 ... there are more ways of plugging your decks in than I can describe in this book. Here's the rule – check the manual against the power system of the country (and the club!), and be sure to take the right adaptors. Remember, plug adaptors are NOT voltage converters.

One important rule that most people don't consider: if you play a club where the power system is dodgy – any club in Ibiza, Ayia Napa, Bangkok, etc., or even some wee American towns – spikes and surges will appear in the power system that could destroy your kit. I once lost three laptops in one night to power surges in a club in Crete. If you're taking your own kit, consider taking a surge protector.

Audio

In most cases, when you arrive in the club you won't have to make any connections that have anything to do with audio. In fact, you'll have a hard time getting the club to let you plug in anything. The club has an expensive sound system they want to protect, so the last thing they need is you plugging in some new toy that will blow their woofers.

Some even take the draconian measure of bolting the mixer into a "well" on the desk. If that's the case, the only thing you'll be connecting is your DVD into the in-house DVJ.

Other clubs are more open, in which case you'll be allowed to plug in your own decks, effects and loop machines.

In all cases, if you're carrying your own kit at all, not only should you send a technical rider, you should also put in a phone call to the technician who'll work with you on the night. Telling the promoter is rarely sufficient. Internal communication in clubland is more bureaucratic than most universities. Find out the name of the technician, and be sure you talk to them.

CDJs

SOUND

The audio connection for all the CDJs is straightforward – a simple stereo pair of RCA (phono) cables. Red to red, white to white, on the back of the mixer. Be sure to plug in to a "line in" or CD jack, and not a "phono in" jack – and make sure the switch on the mixer is set to "line" or "CD" and not "phono".

The DVJ-X1 and some of the CDJs also have a digital audio connection. Several mixers have this feature, and if yours does, by all means hook up a digital cable between your decks and the mixer. The sound is sweet!

CONTROL

DJM mixers also have a control cable to auto-start/stop the decks when the crossfader or upfader is moved. This can also be useful for auto-switching the video mixer.

Effects

Effects units like the EFX-500, EFX-1000, C-Loops, etc. are connected in several ways.

SEND/RETURN

Most high-end and mid-range mixers have send and

Promoter power

When it comes to organizing sound in the venue, promoters have almost no control over what the club decides will happen. Promoters are, generally, independent entrepeneurs and club owners despise them. To make matters worse, the door staff despise them even more and, on the night, the door staff are gods.

If you need to know what voltage, equipment or specifications the club operates, ring the manager of the club. Almost all club telephone numbers and emails worldwide can be found at www.mim.dj

---

1. RCA (phono) cables

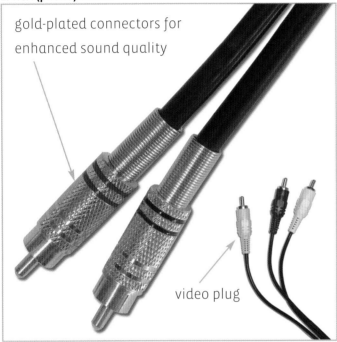

gold-plated connectors for enhanced sound quality

video plug

2. Digital audio connection

3. Control cable

4. 1/4" Jack

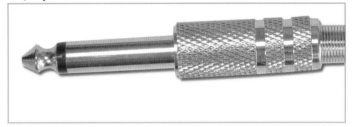

5. The pin of EFX cable

6. Component BNC cable

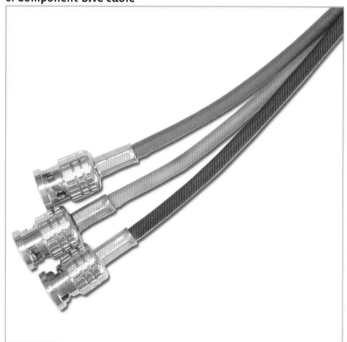

8. Component BNC cable

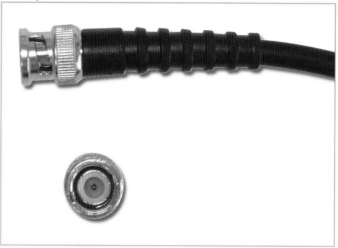

7. S-Video to BNC adaptor

9. S-Video

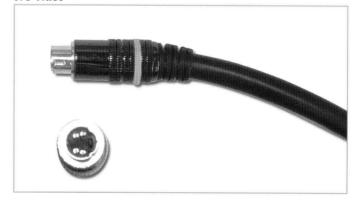

return features. The connection is made either by RCA (phono) cables, or a pair of 1/4" jacks, or by a special digital cable like the one that passes between the Pioneer EFX-1000 and the digital Pioneer DJM mixers.

IN-LINE

An in-line connection is sometimes necessary, particularly when there's no send/return on the mixer. In this case you would use an RCA (phono) cable to connect the output of your deck to the input of the effects unit, and another cable to connect the output of the effects box to the input of the mixer. Again, be sure to use "line in" and not "phono" on the mixer, and set the switches appropriately.

Theoretically, you could connect your effects unit in-line between the mixer and the amp, but there's no way any club will ever let you do this. I don't suggest doing it at home either. The chances of blowing either an amp or speakers are far too high.

CUE JACK

Some effects units take an output from the headphone jack on the mixer – particularly looping and sampling units. You choose which channel goes to the effects box by choosing which channel is in cue. You connect the output of the effects box into an input channel on the mixer, as above.

There will be a loop-through on the effects box that still lets you use your headphones for cueing.

This method is particularly useful if either your mixer lacks send/return features, or if they're being used by an effects unit. Some mixers, however, have more than one send/return channel, like the DJM-1000.

Video
PROGRAM CONNECTIONS

There are four options for connecting the video outputs of your DVJ-X1 to the club's video system, or mixer:

- Component BNC – these connections are marked Y, CB/PB and CR/PR, and are connected using three BNC cables, which lock onto the jacks. It is truly rare to find a club video mixer that accepts component BNC.

- Composite BNC – this connection uses a single BNC cable to the mixer. Clubs with more professional mixers, like the Panasonic line, will often have this type of connector.

- Composite Phono – this is the most common connection to make to a video system, and is made via a single phono (RCA) cable. Video phono cables are of higher quality than audio phono cables, and although they are interchangeable, the quality of video will degrade significantly if you use a cheap audio cable.

- S-Video – this is also a common connection, using a single S-Video connector.

PREVIEW CONNECTIONS

From the DVJ-X1, you connect either a single Composite phono or S-Video connection to your preview monitor. Why use a preview monitor from the DVJ-X1 when many mixers allow you to preview what is coming in through a channel? Because the preview output of the DVJ-X1 is the best place to access many of the advanced setup, cue and loop, and menu features of your DVD decks.

ADAPTORS

Adaptors are a reality when you're connecting video to a club system: BNC to phono, S-Video to phono, S-Video to BNC, etc. Many VJs carry around a toolbox with a selection of adaptors and connecting cables, but truly, for the DVJ, lugging a toolbox into the club is not a good look. I prefer one of those velvet Crown Royal bags with the gold drawstring.

IN THE CLUB

THE FIRST TIME A BEDROOM HERO HITS THE CLUB, THEY'RE IN FOR A SURPRISE. LEGS SHAKE, FINGERS TREMBLE, EVERYTHING SOUNDS WRONG AND THEY FLUB EVERY OTHER MIX. IF THEY'RE LUCKY. DOES THAT MEAN THEY SUCK? NOPE, JUST MEANS THAT SOMEWHERE BETWEEN THE REHEARSAL SPACE AND THE DANCEFLOOR, ALL THE RULES CHANGED.

Sonic shock

This gets everyone at least once in their DVJ life. You walk into the club, whack on your first tune and everything sounds ... wrong. Too much bass, not enough treble, you can't hear well enough to mix, or things sound out of whack.

The sonic environment of a club is the worst possible environment for mixing tunes. Mid-range and highs pumping out from the sound system slap off the back wall and come back on the DVJ about half a beat behind everything else. The deeper the room, the greater the effect, and in some large halls, you can almost estimate the depth of the room by counting beats the way you estimate lightning against thunder.

The effect: when you're mixing to what you hear in the room, you're actually mixing in at least half a beat behind what the dancefloor is hearing. Disaster.

It's enough to shake any fledgling DVJ, and even some with experience – and believe me, if there's a solution the venue can provide but that hasn't materialized by the time you arrive, they're not going to understand when you demand it.

MONITORS

Good monitors in the booth are a godsend. By cranking up the monitors, you can cover the effect of beat slapback in the room and get your mixes on target. But decent monitors are also the last place the promoters will spend money. Poor monitors in quality clubs are commonplace and are a DVJ's worst enemy.

Many can't give you the volume you need to cancel out the dancefloor boom without distorting. Distortion not only causes deafness in the long term, it also wears out your ears over the course of a set.

There are two easy solutions. First, turn down the monitors between mixes. This may not be much fun, but it will help your set and save you from becoming an old man who turns the television up so loud the neighbours complain.

HEADPHONES

The second solution is to use cue/monitor mixing in your headphones. With cue/monitor mixing, you can compare in your headphones what's in cue with what's going out on the floor. And by mixing between the two, you can easily go from total cue to total master sound.

Many DVJs hate this way of mixing. It's an extra knob to twiddle when you're in the mix – and if you're already handling cueing, EQ, upfaders and a video mixer, one extra knob is one too many. It also makes it more tricky to handle your levels: if you forget to dial over to the house mix in your headphones, then you'll likely think the tune you're bringing in is louder than it really is, killing the volume and the vibe.

With practice, though, this type of mixing is ideal. First, it eliminates the sonic shock you experience between the studio and the dancefloor. Second, it means you'll never have your set sunk by cheap monitors again. Finally, keeping the monitors either off or low, and the volume relatively low in your headphones, can stave off DVJ deafness for at least a few more years.

What the other DVJ thinks?

You're warming up for your DVJ hero. But he's just standing behind you, chatting away with the promoter. No smiles when you perform that perfect scratch routine. A bit of a frown when you mix from one tune to another. "Oh my god," you panic, "they all

[The answer is to DVJ like an Italian driver –
pay absolutely no attention to what's behind you.]

think I'm terrible!"

Every DVJ experiences this almost every gig. The answer is to DVJ like an Italian driver – pay absolutely no attention to what's behind you. There's only you and the dancefloor.

You may imagine they're standing behind you thinking, "This guy's rubbish! How did he get booked?" But he may just as well be thinking, "This guy's good! How am I gonna follow him?"

Guess what? You'll never know.

When it all goes horribly wrong

Q: What do you do when ...

   ... you botch a mix?

... they hate the tune you just dropped?

... someone's leaning over your decks and frowns at what you've just done?

... someone is jabbering in your ear and won't get out of the DJ booth?

... one of the decks goes wobbly?

... when the promoter tells you this is the last gig you'll ever play?

... you're too tired to carry on?

A: Smile, nod your head and drop another tune. The only way to the end of the show is through it.

Sex

Playing a great DVJ set is like making love to a roomful of people en masse. You have to get their attention, seduce them into your space, take them through foreplay, climaxes, cool-downs, and climaxes again. And you've gotta make it last all night long.

Reading the crowd

With a lover, you won't know exactly what to do with them until you get them into bed. Once you're there, you can exercise your tried and tested best and see what they like. In some cases, you'll make up new moves on the fly. But a great lover will always watch and listen in order to know what works and what doesn't.

The same happens with a dancefloor. You may have an idea of what you'd like to do, and you may have some of your routines down flat, but the great dancefloor lover will watch the dancefloor as much as the decks – to see what works, to feel where the energy lies in the crowd, and to respond with their best moves.

Don't worry, you're still in control, but the dancefloor is also telling you how it wants to be controlled. Do it wrong and this is your last date. Do it right and they won't let you go home.

Pacing

It's the rare dancefloor that will be packed all night long. Watching other DVJ sets at festivals or mega-clubs leads you to believe everybody's dancing all the time. Big dance events have an ever-changing crowd, though, and when punters upfront are tired of dancing, they move to the back and the next batch comes forward. It's an endless stream of people ready to put their hands in the air.

Most gigs work differently. You're only playing a couple of hours, but the dancers are there all night long. They want to be driven hard at times, but they also need the occasional bit of tenderness and maybe even a breather.

A great DVJ spots when the crowd is ready for a break, and brings the pace back down. Once they've had a rest, they'll be ready for more and want to go even higher.

Give and take

DVJing isn't just about playing great tracks, it's also about setting up an exchange with the audience.

Sometimes it's about you and sometimes it's about them. Moving the attention back and forth between the DVJ and the dancefloor is a skill few DVJs can manage, but the mechanism is simple.

At some point in the set, the DVJ grabs the audience's attention. You can do this with a short bit of scratching, a little EQ flair, a truly unique remix, or great effects across a build-up. You can tell you have them when they all turn to face you. What do you do then?

[Stuart Warren Hill – Hexstatic]
The DVJs are brilliant because you can DJ CDs with them, VJ with them or perform a spectacular AV show with them using DVDs. There are so many things that you can do content wise, but the main thing is to try and be original.

James Mitchell

Smile, nod your head and give it right back to the audience, with a boom! Drop a groove. They'll cheer, shake your hand, and then get their heads down and dance. Pretty soon, it'll be time to take it back again.

Riding these waves, building tension and giving release is the essence of creating a great DVJ set for the dancefloor. And the shape of your set is the shape of you as a DVJ. Develop your own style. Danny Tenaglia is famous for his marathon sets. Jeff Mills is known for taking his dancefloor on a journey.

Lead them whatever way works for you.

The through the roof set

(maximum length – three hours)

This is the way most festival DVJs, and new DVJs, go at it. Start low and build to a peak, like going up a ramp. The set rises across the night and finishes with a bang. This works best for trance, hardcore and other hands-in-the-air styles, and for sets under three hours.

[Robin Brunson – Hexstatic]
From my point of view and coming from a DJ background, this was always the way I thought it would go ... For us, with DVJs we can pretty much do everything we ever wanted to with video.

If you're warming up, be careful. You only want to go as high as where your next DVJ will be about 20 minutes into their own set (see "the warm-up set").

The sine wave set

This set rolls along like waves out at sea, with peaks and troughs building the audience then giving them some space all night long.

The sawtooth set

Like the sine wave set but closer to the shore, the sawtooth set builds peaks and troughs, but the high and lows are sharper. European continentals and big American rave audiences often love this, with deep breakdowns, huge build-ups and massive peaks.

The hybrid through the roof set

This is my favorite: synthesize the sawtooth or sine wave with the through the roof set, and it's like the best kind of sex you've ever had – just make sure there's a massive climax at the end. And this one can

carry your through five or six hours at a session.

The warm-up set (early)

The warm-up DVJ has the hardest job in the club – getting a roomful of people-who-want-to-dance to actually dance. It doesn't sound like much, but oh-my-god-is-it-difficult!

The warm-up DVJ is the ultimate seducer. You gently coax and whisper your dancers out onto the dancefloor one at a time, watching all the while to see if they're responding. Two or three girls might love it when you drop an old Laidback or Prince tune. That's your start. Let the groove roll along, then bring in another classic. Now you've got a few more. And a few more. Before you know it, the whole house is rocking.

But too big or too hard, too soon, and you'll scare them off before you get started.

The warm-up set part II (early-ish)

This is easier than the warm-up set, in that you should already have a few dancers on the floor from the previous DVJ. But it's considerably more difficult in that now you must consider the politics of the dancefloor. And often the first warm-up DVJ got it wrong, either going too hard too soon, completely failing to interest the dancers at all, or playing all the best tunes and leaving you with squat.

First warm-up sets tend to be funkier and slower. Your job now is to take the music up a level, get the audience really ready for a party, but only deliver so much. Why? Because the headliners come on next. They're the one who'll get to have all the fun.

Steal their thunder, or play their tunes, or take the dancefloor too high too early and you'll never be playing in this club again.

The post-peak set

Imagine you're following Fergie at 5 am and you'll get an idea of the post-peak set. This is a place where you can't possibly take the dancefloor any further than they've already gone. They're ringing with sweat and ready for a cool-down.

The post-peak set is a great place to do one of two things: experiment with a darker shade of dance music or pull out the old classics. Either way, this set will be all about the stragglers.

The VIP set

The VIP room is the worst place in the club, and having to play a VIP DVJ set is either a cruel initiation ritual or some sort of severe punishment for arrogance in a former life. The people in the room are either the promoters, other DVJs, industry-types or hyper-self-conscious blaggers who are so worried about getting ejected they can't be asked to dance. Playing in the VIP is like playing an all-night warm-up set in a half-empty diner. How do you handle it? Keep it funky (the VIPs are a little older than the rest of the club) and keep it simple (see "keep it funky"). And remember, if you get the VIP set to dance, you'll be in the main room next time.

The high street club set

AKA, trotting out the hits for the high street girls. In America, I called this "playing the Ramada". The high street DVJ has two priorities, in reverse order:

- Keeping the dancefloor filled

- Clearing the dancefloor occasionally, so they all go to the bar and order drinks.

Clubland is about drink sales, and nowhere more so than in a high street, chain or hotel nightclub. High street or Ramada DVJs will even use slow dance tunes to get the drinkers to the bar. Blech!

The eclectic set

This is my other favorite type of set, and is an essential part of the marathon set. The eclectic set is simply about playing a variety of music. At its best, the set is described as a particular genre, with a hint of something else: Breakbeat with a bit of tech house; R'n'B with a smattering of hip-hop; House with a bit of breaks and acid.

Erick Morillo is known for dropping a rock track into his sets. This works well, but drop too many and you'll lose your audience. The eclectic set is about keeping the groove rolling, but breaking it up with a rare, unique and totally unexpected moment. Mix it up too much and you're be playing pastiche.

Club promoters hate eclectic sets, as it makes it harder to sell the night to a specific crowd, but some punters love them. They listen to a dozen kinds of music on their iPod, so why shouldn't they dance to them as well.

The terrace set (with sunset sizzle)

The terrace set (AKA the chillout set) is an Ibizan specialty. If dance music were yoga, the terrace set would be the Hatha variety. Chilled tracks beautifully blended with a smattering of classic vocal tunes.

Sunset sizzle

Mambo in San Antonio (that's Ibiza, not Texas) is known for its amazing sunset terrace sets, with a special sizzling accent reserved for the moment the sun touches the horizon.

The marathon set – one DVJ, three "record boxes"

Presented with the prospect of a 12-hour set, most DVJs panic. Then they work hard for a month figuring out how to do it. It's simple:

- Treat the marathon set like any other set, just pretend you are three DVJs. Take three boxes of tunes. Take three different styles.

- Avoid booze and drugs. You'll never make it through to the end. And bring some extra-long tracks so you can make a dash for the loo.

DVJ MASTER

MAKING YOUR OWN VIDEO IS ONE OF THE FUNDAMENTAL PLEASURES OF DVJING. WHETHER YOU'RE CREATING ABSTRACT CLUB-STYLEE PIECES, MASH-UPS OF THE LATEST MTV OFFERING, OR ORIGINALLY SHOT WORKS WITH YOUR FRIENDS. CREATING YOUR OWN VIDEO WORK INFINITELY EXPANDS THE CREATIVE PALETTE OF MUSICIANS AND DJS ALIKE.

Samples and originality

Samples are great, and the history of sampling goes back further than you think. Beethoven used to sample passages from other composers' work. So much so that he gave one of the greatest creativity quotes ever rendered:

There is brilliance in imitation, but genius in theft.

I don't quite think he could have foreseen the future of direct digital sampling.

True, if you're a sampling master, you can create a unique sound out of any sample, no matter how often the source has been used.

More likely, you'll end up using samples that are available to everyone, as well as ones that are on everyone's mind – meaning you're going to sound like everyone else.

The better route is to nick a musical idea and then record it yourself. Not only will you not have to pay performance rights, you'll also create a sound that no one else has created. Do this enough, and you'll achieve the holy grail of all producers and composers: you'll find your sound!

The same applies to video. And with video, you also need to consider the context of the club when you select and use video samples. Sure, it's fine to sample a moment when Jessica Alba turns her head toward the camera with lust in her eyes. But get in the club

[There is brilliance in imitation, but genius in theft.]

and play this out, and the punters are going to be thinking: Hey! It's Jessica Alba! What the heck does she have to do with this tune?

Much better is to put that shot in your head and then re-shoot it with a friend in Jessica 's place.

The camera

This is the weapon of choice for creating your look as a DVJ. You may think you're just an average music-loving DVD-playing punter, but truth is, no one sees the world quite the way you do.

Whether you've had an unusual childhood, have a peculiar color taste, prefer to look at people from a specific angle or you're near-sighted, you have a particular view of the world. And you can translate that to the screen with your average handycam.

Volumes have been written over Cézanne and Renoir's blurred, somewhat abstract paintings. Guess what? They were short-sighted. Although they did much as artists and thinkers about the form their art took, myopia in no small way contributed to the development of Impressionist painting. They painted their world from their perspective, as you should shoot your world from your particular point of view.

Carry your camera everywhere you go, charged and loaded with tape. Whip it out whenever something inspires you. And shoot, shoot, shoot.

Two hours' shooting on a charged dancefloor, an hour with your friends at the beach, 30 minutes in an urban industrial wasteland – put them all together and you'll have enough footage for more than one three-hour set.

One of the keys to remember: DVJ video isn't like television or film or even MTV! The principal difference is you can repeat your video content and still keep your audience riveted.

Dance music, whether House or Hillbilly, is based on repetition, as is nearly all music. Themes repeat themselves – sometimes straight, sometimes with variations, sometimes in sequence, and sometimes across the course of a two-hour mega-symphony.

Music video in the club does the same. In fact, if you play a three-hour set without repetition, you'll have your audience riveted to the screen, rather than firmly planted on the dancefloor. If you want to make a movie, by all means, write a script and get to it; I'll see you in a couple of years. But if you want to create a dynamic DVJ set, let yourself get involved in repetition. Don't worry about narrative; shoot everything you see.

The kit you'll need

What to look for in a video camera and in a computer for editing video. In this example I use an Apple PowerBook and *iLife*, but you can just as easily choose a PC laptop (or desktop) and another suite of software. I also have a look at DVD recorders.

Point. Shoot. Edit. Burn. And you're there.

The camera is, without a doubt, the fastest, cheapest and easiest way to create video, and create it in your own style.

Software

Software is another viable alternative to shooting your own video. There are dozens of programs for creating video-based content, as well as programs for creating your own DVDs. There are even more books designed specifically to teach you how to use these programs. I'll cover the easiest to use, because I think you should spend most of your time concentrating on performance, but I'll also give you enough information to get you through the process of creating some basic DVDs, and introduce you to some categories of program that will be useful if you want to go deep with your video creation.

ANIMATION SOFTWARE

If you're particularly plugged in to graphics, characters and drawing, animation could be the answer to your dreams. If you're too shy to gather your friends, family and a few attractive strangers for a shoot in the woods or down a gritty alley, animation is another way of getting video content up on the screen.

There are three fairly easily-learned packages for creating animated graphics: *After Effects* from Adobe, and *Flash* and *Director* from Macromedia (now also owned by Adobe). When I say easy, I mean it, but keep in mind, it will still take a couple of months to get to grips with any animation package.

Flash and vector graphics

Flash is specifically designed for manipulating vector graphics, which allow you to create small, sophisticated animation files that are easily downloaded from the web. Vector graphics are based on graphical shapes like circles and squares. Vector graphics work like MIDI recording: when you store a vector graphic image, you store a mathematical description of an image, rather than the actual image itself; when you record MIDI information, you record information about what the sound sounds like, rather than the sound itself. In both cases you create a small file.

An animated vector graphic of a circle crossing a screen works like this. The size, color, outline and position on

screen of the circle are recorded as data. Then, another frame, or moment, of the animation is recorded, say a second later. This second frame in this example would be of another position for the circle. When the file is played back, the software reads the information about the circle and draws it. It then looks ahead to where the circle will be in a second, and redraws the circle according to the data about size and color, etc., several times across the course of a second, creating the impression of a circle moving across the screen.

Vector vs. bitmap

vector

Some of the postcript code used to describe the vector circle

```
/lin§fg 0 get 0 ne§l currentpoint 0 dop m§
§currentpoint/@3 X/@4 X g np/@1 X/@2 X
fp§@4 @2 lt§@3 @1 ge§@4 @3 m @2 @1 l pnsh 0 rl
0 pnsv rl @4 pnsh add @3 pnsv add l pnsh neg 0 rl§
§@4 @3 m pnsh 0 rl @2 pnsh add @1 l 0 pnsv rl
pnsh neg 0 rl @4 @3 pnsv add l§ifelse§§@3 @1 gt
§@2 @1 m pnsh 0 rl @4 pnsh add @3 l 0 pnsv rl
pnsh neg 0 rl @2 @1 pnsv add l§§@4 @3 m pnsh 0 rl
```

Handle bars used to create bezier curves

bitmap

A circle made of indiviual pixels

Combine several circles, squares and lines, and you have a drawing.

Director and bitmaps

Bitmapped images are recorded differently than vector graphics. In a bitmap, a grid is metaphorically placed across the image, and information is recorded for the color and luminance, etc. of every visible bit. Multiply this times 29.97 frames per second, the frame rate of an American video, and you end up with a huge file. Not good for downloading over the internet, but fantastic for creating animations.

Director is specifically designed for creating both bitmap and vector-based animations with interactivity.

Adobe After Effects

After Effects is designed specifically for manipulating video images. You can take any video you've shot, get it in your computer and manipulate it in almost any way imaginable. You can also create bitmapped graphics – like logos – and animate them.

The beauty of *After Effects* is that the program itself also lets you use plug-in filters for manipulating video, making *AE* infinitely expandable (but infinitely expensive).

Synthetik Studio Artist 3.5

This phenomenal piece of work is the only majojr piece of software available capable of painting, drawing and rotoscoping video automatically. Any still image or video file plugged into *Studio Artist* comes out as an instant animation, in any one of thousands of pre-built (and tweakable) styles. Not only this, but *Studio Artist* can also automatically turn video files into vector-based (like Flash) animations – that means big stylish video at very small sizes should you want to load work up on the Net.

There are two potential approaches to *Synthetik*

Studio Artist for new DVJs – first, use it fully on automatic, and create an astonishing range of video effects automatically; or second, bury yourself in this program and become a true master of "liquid media" video. The caveats? First, it's only available for the Mac. But with the new dual-core Intel processors, Macs have more than enough number-crunching headroom to handle video renders. Second, there are more than 8 hours of training videos – a benefit if you've chosen to become a *Studio Artist* master, but a bit intimidating if you are tight for time.

I think you should view it as a benefit. *Studio Artist* is so capable as an automated effects processor, I'm sure you'll find it suitable no matter the level you choose to approach it. Not to be overly effusive, but I choose it hands down over any other render-based video manipulation software – worth shifting to Apple for.

GENERATIVE SOFTWARE

Generative software for video rarely produces results more interesting than a cheesy screensaver.

There are several programs of this type, from eye-candy toys like *PixelToy* through to professional generative packages like *Artmatic Pro*. These programs let you set a few parameters and press go. What they make is abstract video that is typically trippy, glowy and, erm, abstract. They're incredibly easy to use, and therefore the video they automatically generate is so common even you will be sick of it after a day or two.

Don't be fooled by simplicity, though. Both *PixelToy* and *Artmatic Pro* allow you to import stills and movies, respectively, and apply effects parameters. This is not only effective for creating animated logos from still graphics, but also for applying effects like flames to a still photograph.

These programs are great for the odd video accent, but playing an entire set of this type of video is roughly the equivalent of playing a whole DJ set with nothing more than a drum machine. I suggest their use as production tools lest you find yourself limited by the creative potential of your software.

EDITING SOFTWARE

Video editing software comes in two varieties: performance VJ software and non-linear, non-real-time editing software.

VJ software

VJ software is stunning for creating video for your sets. Most VJ software not only lets you fire off video clips in real-time from either your music or computer keyboard, it also lets you apply effects to the video. VJ software is generally designed for taking a laptop to the club and playing a mute visual set, firing off clips in time to the DJ's tunes. However, we're DVJs, and not really interested in playing second string.

Nonetheless, VJ software is one of your most effective tools in the studio for creating music DVDs. You can load up a single on a CD player, and "play" video along to your track, recording it and creating a DVD track to play out in the club later. VJ software is also useful for creating text graphics and logos on the fly, and playing them back live in clubs.

The best VJ software, like *Grid Pro, Resolume* or *FlowMotion*, also let you apply effects to your video clips and record your performance on the computer.

Non-linear editing software

These are straight-ahead editing programs, and come as simple as Apple's *iMovie*, or as complex as Apple's *Final Cut Pro* or Adobe *Premiere*.

Editing software does what it says on the box. You hook your camera up to your computer through your Firewire port, record the video onto your hard drive,

Firewire – 6-pin computer to 4-pin camera

drag the video into a timeline, and start cutting and pasting a music video.

iMovie is so simple to use, you'll never need a manual. *Final Cut Pro* comes with FOUR manuals. Both take plug-ins for creating effects, both have fairly sophisticated handling of audio, both handle HD, and both are miles more advanced than the big video editing suites your dad had to use.

One more thing: *iMovie* is free on every new Mac. *Final Cut Pro* and *Premiere* definitely are not. The choice is yours.

LEARNING SOFTWARE

One last word on the range of software: as with guitars, drums and decks, maintaining your software chops requires practice and constant update. Take

your average DVJ – he wants to create his own DVD music videos, produce his own killer cut-ups, do the cover artwork for his bootleg mix releases, maintain a website, a blog and a fan forum, and get his tracks up online to download. Plus there's *Excel* for gig schedules, *Word* for contracts and press releases. Add to that a steady stream of software updates, and most DVJs will find themselves so overwhelmed with poring over software manuals they have little time to work on their performance skills.

The answer? Lynda.com

Lynda.com provides hundreds of hours of online software training, in streaming video format, for less than a tenner a month. Whatever new pro package you need to get up to speed on, Lynda.com has a set of tutorials. Although you'll find little there in the way of VJ or MIDI sequencer software training, you'll find everything you need to know about most of the major graphics and web packages. I discovered it while writing this book and now use it almost every day.

DVJ MASTER

CREATING YOUR OWN VIDEO ISN'T THE ONLY WAY TO GET CONTENT FOR YOUR DVJ DVDS.
As I said, there's genius in theft. Particularly when you've paid for what you're stealing and there's no chance of getting sued. Here's how to find libraries heaving with content, ready for playing out.

Stock services

Stock services provide video and photographs for designers, artists and DVJs(!), on tape, DVD and as computer data. Prices vary from free to way out of your league.

For non-commercial use your top source is the BBC. In 2004, the BBC announced the placement of video content from the massive BBC library online for domestic public use. This means, technically, you can't use it to make money. And you can't use it if you live outside the UK. What I would suggest is that you can't use it to create a music video that you would release for sale. However, I don't see how they could

object to your creating a music video for personal club use within Britain. What that means, I won't say. Check with your lawyer and your local immigration official.

For commercial use, libraries like Corbis lie at the other end of the spectrum. Corbis sells still and motion images to ad agencies principally, but anyone with deep pockets can buy in.

You'll find both online.

VJ content

Several VJs have begun making their content available online and through mail-order. VJs are great sources for club-oriented video content. They've been making club video for decades and know the audience.

Your first and best source is to scan www.vjforums.com and its sister site

A list of content resources

Dandelion Collective – www.dandelion.org
Dandelion Collective are a group of artists, VJs and motion graphics designers based in Vienna who have been selling DVJ visual content on DVDs for years. Their DVDs are inexpensive, usually half an hour long, come in PAL, NTSC and digital format, and contain a range of visuals built to a theme. Their latest offerings are cut to a specific bpm, with a kickdrum and hi-hat track recorded underneath. This let's you sync your visuals to the beat by simply beatmatching the DVD using the headphone cue. Your first source of content for the club – highly recommended.

Creative Archive – creativearchive.bbc.co.uk
Greg Dyke is one of the great minds of his generation, and one of the most important people in British broadcasting; perhaps the greatest legacy from his time at the BBC was the founding of what became the Creative Archive License Group. The BBC, Channel 4, the British Film Institute and Open University got together, did the paperwork and started putting their archives online. As they say, "Find it, rip it, mix it, share it, come and get it!" This is gonna change AV creativity forever – just be sure to read the licensing info carefully.

Creative Commons – www.CreativeCommons.org
Creative Commons is an effort to introduce new copyright rules for the creative public, and to make tracks, video, photos and text available under a variety of licensing schemes, including free. What matters to you is the thousands of films available online for free. And the tunes!

BBC Motion Gallery – www.BBCMotionGallery.com
Encompassing huge portions of the CBS and BBC archives, BBC Motion Gallery gives you access to more than 300,000 hours of footage online and more than a million hours of offline footage. Could you

A list of content resources – cont.

ever need anything else? Yes, a fat wallet.

Corbis – www.corbis.com
The granddaddy of all image archives. They may be too expensive for most DVJs, but if you're looking for a particular shot and can't find it on Corbis, chances are you'll have to shoot it yourself.

Google – www.google.com
Type "stock video" into Google and you'll get more than 100,000 hits. Somewhere in there is exactly the video you need, at exactly the right price. (I particularly enjoy Teton Gravity Research's videos of insane boarding heroes.) Just be sure to check the license.

www.vjcentral.com. VJForums is the number one information source for VJs. My website www.howtodvj.com has a comprehensive list of up-to-date DVJ sites.

Promo Only and Mixmash

Promo Only and *Mixmash*, as well as a few other subscription-based companies, sell DVDs and pre-made music videos specifically for DVJs. My argument *against* these services is that you'll be playing out the same content as every other DVJ. My argument *for* them is that they're relatively cheap, time-efficient (vis-à-vis creating loads of your own content) and provide video in a format that even the most wet-behind-the-ears beginner can understand. Qualitative judgements are up to you, but I'd add this: using generic video club loops is somewhat like using generic drum loops in your tracks. Without some degree of manipulation, your work will not only sound (or look) like everyone else's, your work will look and sound like the lowest common denominator. Just as you can buy one of these videos and use it in your set, so too can the Hilton use it as a backdrop for their lounge singers. Think of these videos as source material, not finished product.

Students, friends and family

This is my favorite source for video content, outside shooting my own.

Does your father still have that old 8 mm film he shot on his safari to Africa? Didn't your mate just get back from boarding in Vale? What about those old videos of you and your girlfriend?

University film and art students are also a great source for video content. Post a notice at your local art school, tell them what you want and how you want to use it, and chances are you'll find yourself with loads of content for free.

DVJ MASTER

I TOUCHED ON VIDEO FORMATS EARLIER IN CHAPTER 4, BUT IF YOU REALLY WANT TO UNDERSTAND WHAT'S GOING ON WHEN YOU ROCK UP TO THE DVJ DECKS, THERE'S A LOT MORE INFORMATION TO ABSORB.

These are a few terms you need to understand just to grock the other terms. They all have to do with the way video is put together. In the old days of film, moving images were fairly easy to understand, particularly because they had physical qualities. "35 mm film format" meant you were shooting on film 35 mm wide. And 24 frames per second meant how you would see 24 frames each second. Doh!

With the advent of video, and image data, opaque to the naked eye, everything changed. Resolutions became horizontal, frame rates went wobbly, and helical scanning was introduced as a complex system of angular frames-not-frames on a video tape. Confused? That's nothing. Digital video pulled any

sense of secure ground out from under the average non-computer scientist's feet.

Here, I'm going to limit the concepts to a few I think are not only essential, but also relatively easy to understand. For a thorough technical guide, check out *How Video Works* by Diana Weynand and Marcus Weise (www.focalpress.com). Use this section either as a reference section or an aid to understanding the material later:

Interlace

A video screen is made up of a series of horizontal lines of color and brightness information. For example, an NTSC (American) screen has 480 horizontal lines running across it. Interlacing is the sequence in which the lines are drawn. In an interlaced image, the 240 even-numbered lines are drawn from top to bottom first, then the 240 odd-numbered lines. Drawing both fields takes about 1/30th of a second on American television (with a frame rate of 29.97 frames per second). Each field, odd or even, takes about 1/60th of a second to draw.

LCD and plasma monitors automatically de-interlace the image, weaving or blending the lines together. This makes the image look more like it was shot with film.

Progressive scanning
Progressive scanning draws all the horizontal lines of a video image from top to bottom. Progressively scanned images look more smooth to the eye, particularly when the camera pans from left to right

or vice versa. Your DVJ-X1 can play progressive video.

Frames per second
A video or film image is really just a series of stills presented to you quickly so it looks like motion. This is how the eye works, taking a series of still pictures and putting them back together in sequence in the brain (called "persistence of vision"). NTSC takes 29.97 images per second or, rather, it is rated at 29.97 frames per second. PAL uses 25 frames per second. 35 mm film, which is used to shoot most major motion pictures, operates at 24 frames per second.

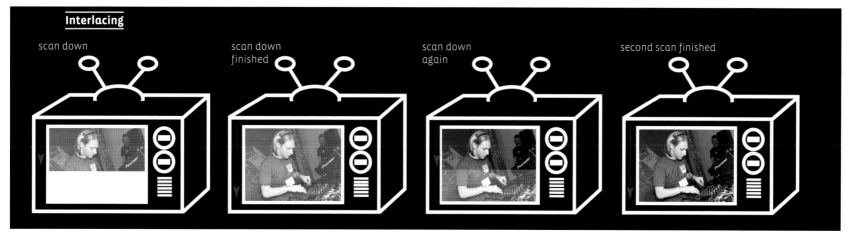

Interlacing

scan down / scan down finished / scan down again / second scan finished

Horizontal resolution

For broadcast video, this is the number of scan lines across the screen. NTSC has 480 visible lines and PAL has 625.

RGB

There are only three colors in an analog broadcast image – red, green and blue. Each dot of color is represented by one of the little dots on your TV screen (get a magnifying glass and have a look). To make a color like purple, a red and a blue dot are placed very close together, and your eye will see it as purple.

Broadcast formats

NTSC

NTSC is the video format for most of the Americas and Japan. It is an interlaced signal with 480 lines of resolution, operating at 29.97 frames per second.

PAL

This is used most everywhere else, except France, the ex-Soviet bloc and much of Africa. It is also an interlaced signal with 625 lines of resolution, operating at 25 frames per second.

SECAM

SECAM is still widely used in broadcast, but rarely in video (tapes, DVDs, etc.). It's also not worth your time worrying about.

Aspect ratios

4:3

This is one possible ratio of an image's width to its height. It's also the original standard aspect ratio for television, video and computer monitors. Not only that, it was the aspect ratio that was used in many silent Hollywood films (expressed as "1.33:1"). When television stations started broadcasting, Hollywood introduced widescreen formats to compete, like 16:9.

16:9

16:9 is the aspect ratio of HDTV (High Definition Television), and also for lots of new computer monitors and new-generation flatscreen televisions.

Aspect ratios

Old skool – 4:3

Brand new skool – 16:9

It's the standard aspect ratio for all programs broadcast by technically progressive networks like the BBC (British Broadcasting Corporation). Most camcorders will now shoot 16:9, your DVJ-X1 will play 16:9, and I suggest you start shooting all your video in this format.

Digital video

Digital video formats can be any of the above, any combination of the above, or just about any damn thing you want to make them. Because digital video playback is tied more to software than hardware (with the exception of HD), programs like *Quicktime* Player are designed to handle all sizes and shapes of video.

I suggest that if you produce digital video for playback in the club, you keep it to the standards of broadcast television. This will make your life considerably easier once you start transferring your DVDs.

UNLESS YOU PLAN ON STICKING WITH THE TOP 40, A COMPUTER IS AS ESSENTIAL TO YOUR DVJ PRACTICE AS YOUR DECKS. USING THE RIGHT SOFTWARE IS CRITICAL, WHETHER YOU'RE PRODUCING YOUR OWN TRACKS, MAKING YOUR OWN VIDEOS OR GETTING BUSY WITH THE REMIXES. SOFTWARE ALSO MAKES IT EASY TO CREATE LOGOS, MOTION TEXT ANIMATIONS AND MOTION GRAPHICS.

Music production software

THE BIG FIVE
Pro Tools – www.protools.com, *Cubase* – www.steinberg.net, *Logic Pro* – www.apple.com

Pro Tools, *Cubase* and *Logic Pro* are all the professional producer will ever need in a one-stop MIDI software-based production studio. All offer multi-track MIDI and audio recording, plug-in software synths, samplers and effects, basic video editing, and compatibility with just about every MIDI instrument and controller available.

Trawling the internet to decide which is right for you should provide you with endless hours of amusement, reading flamey debates between software nerds who've forgotten to take their medication. For most DVJs and dance music producers, all are starting to look big and unwieldy. More than a few hit tracks have been produced using dance-oriented software like *Reason* and the new *Ableton Live*.

Reason – www.propellerheads.se

Reason was one of the first pieces of integrated studio software to make a difference. *Reason* gives you sequencing, sampling, synths and effects all in one piece of software. The interface is beautiful, and involves pretending to do things like crawl under the decks and hook them up with cables (I'm only half kidding), and should make sense to even the most technophobic DVJ.

Ableton Live

Ableton Live is taking a radically different approach to music creation, *Live* started life as a sample-based composition tool. Now all grown up and ready to go, *Live 6* features MIDI sequencing, not only of internal instruments, but also for all those shiny music boxes in your studio. Even better, Ableton Live now supports VST software plug-ins, just like *Cubase, Pro Tools* and *Logic Pro*.

Most important, however, is *Ableton Live's* ability to place a grid across the beats of any track, and bend it into march-step or groove quantized shape. So, should you have an acoustic guitar track played by someone who can't keep to the tempo, *Ableton Live* will quickly analyze the track and place your guitarist's downbeats where they belong. Translated for DVJs, this means *Ableton Live* can instantly beatmatch any two tracks (almost – actually, most tracks need a bit of manual tweaking in the studio before you take them out to the club).

Many die-hard fans of the more complex systems have dumped their old software and gone over to *Ableton Live*. And because it features context-sensitive explanations of all the functions, available on-screen all the time, the learning curve is nearly as short as for the DVJs. I use it every day and can't recommend it highly enough.

Pioneer's DJS

In an extraordinary move, Pioneer have taken most of the features of a DJM mixer and a pair of CDJ decks, and packaged them into a piece of PC laptop software. Not only does *DJS* arrive as a fully-formed piece of rock solid software, it comes with more effects than any other standard DJ software on the market. Auto beatmatch, auto mix, auto everything. One laptop equals a whole new pair of decks and a mixer. A great production and performance tool, and a brilliant addition to your standard DJ rig.

Video production software

THE BIG FIVE

Apple *iLife* suite – www.apple.com

As an all-around package for DVJs, nothing comes close to *iLife*. The *iLife* suite is packed with every piece of software you're likely to need as a DVJ:

- *iMovie* – video editing software with effects

Pioneer DJS screen shot

- *iPhoto* – image editing and cataloging software for your digital photos

- *iDVD* – DVD authoring and burning software

- *iTunes* – MP3 playback, cataloging and CD burning software

- *GarageBand* – MIDI, digital audio and loop-based sequencing and sampling package

- *iWeb* – easy web development package to get your images and tunes up online fast.

There was a time when integrated packages like this featured the most mediocre piece of software from every category. Those days are gone. Unless you're producing a feature film, there's very little you can't do with *iLife* suite, whether it's making your own edits of tunes, creating music videos, burning DVJ DVDs, or just listening to tracks on your way to the gig.

iLife is so simple to use, you will never need to crack the manual or buy a book to teach you how to get through it. It's dead easy. That means zero learning curve, which means you can get to work today.

Not only that, I've produced tracks for broadcast on BBC television using nothing more than *iMovie*. If it works for the Beeb, I promise you it'll work in the club.

Did I mention it's free with every new Mac? And upgrades are usually around $80? If you haven't got it, get it – and if you haven't got a Mac, maybe it's time to rethink PCs.

Apple *Final Cut Pro* – www.apple.com

Before *Final Cut Pro, AVID* was the only game in town for professional video editors. Apple changed all of that when Steve Jobs returned to the helm and started pushing the Mac and its software as creative machines.

Final Cut Pro and *Final Cut Express HD* give you every tool you'll need as a video producer, whether you're cutting a commercial for a local television station or creating an offline edit of a feature for transfer to film. *FCP* accepts a huge range of plug-ins for both audio and video, and has some of the most sophisticated audio editing available in a video package.

You can also export straight to *iDVD* or *DVD Studio Pro*, which makes your life easier as a DVJ who wants to do nothing more than create DVJ music videos on DVD.

What's the downside? First, a steep learning curve. It is not obvious how to use *FCP* the first time you start the program, and wading through the four – volume manual is a chore. Second, it ain't cheap.

Adobe Premiere and the Adobe Production Studio

Adobe Premiere was the first industry-standard

desktop video editing system. Every bit as professional a package as *FCP*, it is the PC equivalent. After a few years falling behind in the digital creativity game, Adobe recently introduced the *Adobe Production Studio* in both a standard and pro version. This integrates a number of professional video packages in the same way *iLife* integrates a group of consumer packages.

Adobe Production Studio

- *Adobe Premiere* – video editing

- *Adobe After Effects* – video effects package

- *Adobe Audition* – multi-track audio recording and editing, with full MIDI and VST support

- *Adobe Encore DVD* – professional DVD authoring package

- *Adobe Photoshop* – image editing

- *Adobe Illustrator* – vector-based drawing package.

Adobe After Effects and Apple Motion

After Effects is the preferred package for manipulating video and motion graphics. DJ logos, 3D generated water scenes, spinning cubes of video, old film simulations – name it and you can do it in *After Effects. Apple Motion* is the newcomer on the block, but looks set to challenge Adobe. Unfortunately, Apple has tied DVJs' hands by forcing them to buy *Final Cut Pro* if they want to buy *Apple Motion*. Smart, huh?

Either way, in this category and the other, there are two games in town – Apple and Adobe – and few can challenge either.

VJ-style software
THE BIG FOUR
Grid Pro – www.vidvox.net

Vidvox was originally founded by Johnny Dekam, an American video artist deeply involved in the underground video performance scene. Despite the street-level roots of this incredible piece of software, it is a rock-solid work of programming and unparalleled in the VJ software world, as is Johnny in the VJ performance world.

Grid Pro is designed to fire video clips, mix them and add effects. After four years of solid gigging with *Grid,* and its earlier incarnation *Prophet,* I've never known it to crash or lock up once, which made it the software of choice for VJ applications around the BBC.

FlowMotion

FlowMotion is a great piece of VJ software designed by artists for artists, with an emphasis on real-time performance, improvisation and jamming. *FlowMotion* lets you play video like a musical instrument, and is great in the studio and in the club rigged to a MIDI keyboard. It is truly a groovebox for video. One of its main features is the ability to quickly create beatmatched visuals to any track – ideal for producing DVJ DVD singles.

Using any DJ track you can match the visual's tempo to the beat of the music by tapping the space bar. Working with MIDI gear you can use MIDI Clock to sync visuals to your tracks in a standard MIDI sequencer.

FlowMotion is specifically designed to allow you to hook up to a DVD recorder for external recording, but my sources tell me a version is on the way to allow you to render to computer disk and manipulate the video in an editor as well. Highly recommended, particularly since it works with both the Mac and PC.

Videoflux LIVE! and *Videoflux EDIT!* – www.videoflux.com

Created by master VJ Micha Klein and his team, the *Videoflux* pair are one of the newest entries on the VJ market – and therefore one of the most sophisticated. Klein sits amongst the world's most successful VJs in my books – he's been at it for more than a decade, and goes from strength to strength. Whether exhibiting at Mary Boone Gallery in New York, coding hot software with his team or touring the planet with Tiesto, Micha leads a life to love.

With *Videoflux*, Klein has stepped up and brought a level of professionalism to the authoring of VJ software that few have managed before him. *Videoflux LIVE!* features an easy interface, an array of effects and a selection of diverse controllers.

Videoflux EDIT! is designed to let you record and edit your performances on a single machine, and then easily output them for import into a DVD authoring package. PC only.

Arkaos – www.arkaos.net

Arkaos has been around the VJ scene for yonks, and along with *motion dive* is the entry point into performance software for most VJs.

The great appeal of *Arkaos* is its cross-platform operability – it works on both PC and Mac – and its ultra-straightforward interface. *Arkaos* let's you drag and drop effects and clips from a bin onto your computer keyboard or MIDI keyboard, and start playing. It is the most straightforward piece of software I've encountered.

Unfortunately, it also tends to crash and hang a lot. I know from experience. At this writing, *Arkaos* is badly in need of a recode. So, while it remains popular with many a beginner VJ, most dump it within a year for a more solid application. PC and Mac.

motion dive – www.digitalstage.net/en/

motion dive is the brainchild of the Tokyo-based VJ Glamoove.

motion dive is one of the easiest packages on the market, but also one of the more frustrating. *md* is based on the two decks and a mixer model. DVJs are given bins of video clips which they can drag into one of two players. These videos can then be layered onto one another in a number of ways.

Although it's easy to use, triggering video in time to the music is nearly impossible. And the software isn't terribly efficient. What *md* is great for, however, is giving you access to on-the-fly animated logo creation, simply by typing in a bit of text. In fact, that is the only thing I use it for. However, it does text so well, I think it's worth buying a copy.

It's a high price to pay for instant text, but I always take a copy whenever I VJ major festivals, just in case an emergency safety message needs to be put up on the screen, or there's a sudden change in DVJ line-up. PC and Mac.

Extra tools you'll need

Autodesk Cleaner – www.autodesk.com

Video conversion package for the Mac and PC. It'll take video in nearly any format and wrench it into nearly any other format. An essential tool.

Quicktime Pro – www.apple.com

If you use a Mac, you want this. For about $30, you can upgrade your *Quicktime player* to *Quicktime Pro*. Why? For 30 bucks, you'll get a video editing/conversion/compression/streaming package, and all you'll have to do is type in the serial number. Not bad.

DVJ MASTER

... OR, HOW TO MAKE A NO-BUDGET MUSIC VIDEO IN 15 MINUTES OR LESS.

Working from scratch, of course, it's not possible. But something equally useful is.

Not too long ago, I was approached by the BBC to create the first nationally telecast live DVJ mix for Glastonbury – 12 hours of television in total. It seemed feasible given the amount of material I had archived from five years on the road, and our initial meeting turned into a signing. As I was leaving the room, happy with our new deal, by way of goodbye the BBC folks said, "Oh, and Charles, it all needs to be in 16:9."

The room went wobbly. All my work, hundreds of hours of video, was in 4:3. I would have to start over from scratch, and the deadline was imminent.

To get through it, my team of Rich Belson and my good friend Deborah Aschheim, a brilliant media

---

How to make your own music video

First I look at how to make a mute DVD:

- Connecting your camera to your laptop
- Importing video into *iMovie*
- Selecting video clips and turning them into loops
- Recording video back to the camera on MiniDV
- Connecting the camera to the DVD recorder
- Dubbing to the recorder hard drive
- Copying the video from the hard drive to a DVD
- Finalizing the disc.

Next I have a look at how to use mute DVDs to quickly make music videos for your choice tracks. Remember, your DVJ-X1s and video mixer are two of your best production tools. I walk you through a complete mix. Presto! Instant music video!

artist who was visiting from LA, created a system to rapidly develop DVJ DVD singles. It took the three of us three weeks, working full-time, to create the base material. Ever since, I've been able to knock out a DVJ DVD single in less than 15 minutes. That means I spend two hours a week working on my set box, adding about eight tunes a week – as many original music videos as most DJs add vinyls.

This book may be about a lot of things, but principally it's about this: how to make a DVJ DVD single in 15 minutes.

Before I get to the technical side, however, I need to look a bit at aesthetics so you can understand why you're doing what you're doing as you're doing it.

MTV vs. DVJ

For DVJs, creating a music video is a different process than for music video directors.

Directors start with a track, conceptualize a video,

TRACK IN with BAD GUY's footsteps

JOHN's head is pulled back by GOON

ZOOM IN on BAD GUY in silhouette

storyboard it, then go into the shooting and editing processes.

A DVJ starts with a track, looks around to see what video he has on hand, and goes straight to the editing and mixing process.

Why?

First, as a DVJ, you're adding at least ten tracks a month to your playlist. You're going to need to work fast – really fast.

Second, MTV music videos are selling a song and selling an artist. DVJs just want you to shake your ass. They're selling the party. This means they can recycle content from old videos on their way out of the playlist, because the broad visual theme of the video is tied to the night, rather than the tune.

Third, DVJ video can and should repeat across the course of the evening, and across different tunes. An endless stream of non-repetitive, quick-cut video puts the emphasis on the screens, sidelining the DVJ, the music and, even worse, sidelining the dancefloor.

Essential kit
REAL-TIME DVD RECORDERS
There are three key methods for making DVJ DVD singles:

- A live mix from DVD decks

- A VJ-style real-time mix from computer, or

- Editing a video on computer and recording it onto DVD

In all three cases there is one piece of equipment I don't think you can do without – a real-time DVD recorder, preferably with a built-in hard drive.

DVD recorders

Time saving is the one overriding benefit of standalone DVD recorders. DVDs use a form of MPEG video, which is a compressed version of normal digital video. Rendering and compressing DVD images on computer using software encoders, even for short videos on fast machines, is often an overnight process. Creating your first crate of DVDs will take you a couple of months full-time, minimum, and for the fast-moving DVJ with a regularly rotating box of video tunes, this just isn't going to work.

Standalone recorders, on the other hand, feature real-time hardware encoding. Dubbing a music video

Pioneer DVR540HXS – 160 GB of hard drive

from either a video mixer or computer is as simple as recording a VHS tape. You'll also avoid the sometimes long learning curves wrapped around DVD authoring, and focus your precious time on perfecting your DVD mixing skills.

A hard-drive model, letting you record video to an internal hard drive and then create multiple copies on DVD, in real time, will double the buy-in price, but save you bundles on DVD media costs when you botch your mix.

DVJ compatibility

You need to make sure the DVD recorder you buy is compatible with your DVJ decks. You also need to both control the input audio levels and record your audio in an uncompressed (PCM) format. Make sure your DVD recorder gives you these options.

One easy solution is to use the same brand recorder as your DVJ deck. I've tested several models of both consumer and professional Pioneer DVD recorders, and found the audio quality excellent and compatibility with the Pioneer DVJ-X1 seamless.

COMPUTER

The other two pieces of essential hardware you'll need are a fast computer with a Firewire (aka iLink or IEEE 1394) port, a stonking hard drive and DVD-R writer for backup (video uses huge data space for both storage and backup), and a video camera with DV (digital video, eh) in and out, again via Firewire.

Both Apple and Sony (PC) computers provide integrated software solutions for audio and video editing, making DVD authoring something any budding porn addict can handle. All new Macs ship with Apple's *iLife*, all the software you'll need to edit your own music videos. Sony has created the *Vegas Movie Studio+DVD* media editing software, which includes *ACID*, a loop-based composing package, as the PC answer to Apple's *iLife*. And don't forget

Adobe's excellent high-end video software suite. Everyone's in the video editing game these days, though, with even shareware video editors commonplace.

How to make that DVD
THE EDITING METHOD

First-time video makers should reach for the easiest tools. This keeps your creativity fresh and your frustration levels low.

On the Mac, that's *iMovie* and *iDVD*, and on the PC, well, it could be anything. Sony's *Vegas* or Adobe's *Production Studio* are the closest PC things to an integrated suite like Apple's *iLife*. Just as *Reason* eliminates the task of integrating several pieces of MIDI software, *iLife*, *Vegas* and *APS* let your video editing program, music production suite, photo librarian and DVD authoring program occupy the same creative space.

Get started

You want to start by importing your tune. If you're importing from a CD or MP3, you're laughing. Most video software handles either format. The ideal is to convert a CD track to AIFF and then import it into your editing program. An alternative is to import an MP3 into the video editing timeline, although if you haven't made the MP3 file yourself, you need to be wary of both sound quality and copyright protection. I recently made a series of DVDs from MP3s which sounded great in a live mix, but the mix couldn't be recorded because copy-protection schemes kept another copy from being made. This is a surprise you don't want when you're recording your Essential Mix debut.

Importing from vinyl

If you're importing from vinyl, you'll need a decent soundcard or audio breakout box, audio editing and recording software, and a professional turntable with a good cartridge. Your audio file, and therefore

your DVD, will only be as strong as the weakest link, and a DVD with audio sourced from vinyl on a cheap turntable is a nightmare to mix in the club.

Treating the audio

After you've recorded your track from vinyl, be sure to normalize the audio levels in your recording software, apply DC offset, edit out any glitches and export the file as AIFF to keep the next steps simple.

Importing video

Once you've imported that tune into your editor, start importing your video. If the video is coming in from a camera, the process is straightforward. A Firewire cable is connected from the computer to the camera before you start the editing program. You should then have controls within your software that allow you to capture sections of the video as you choose.

If you don't have a Firewire port on your computer,

Audio breakout box

you can buy breakout boxes that convert analog video signal to digital data. You can, but I don't recommend it. A better solution is to install a Firewire card if it's a desktop, Firewire to PCMCIA card converter if it's a notebook or, better yet, buy a new computer.

The bin and the timeline

The clips you import will automatically get stored in a folder typically called a "bin". You can then drag them into a timeline alongside the music track, and trim the front and back of each clip to suit your taste. Lots of editing programs offer multiple views of the timeline. In *iMovie*, for example, one view lets you treat the video clips as discrete blocks chained in a line, while another view is time-oriented, allowing precision editing to beats, markers and clock – not altogether different from many MIDI- and audio-editing packages.

iMovie timeline

main screen

clip dragged from the bin onto the timeline

The bin

add effects, stills, music...

timeline

Exporting to DVD

Once your video is complete, you'll want to export it to DVD. If you've chosen the hard road, export your movie as an MPEG-2 file, then import it into your DVD authoring program. Some media suites like *iMovie* will let you export your movie straight to the DVD authoring program, and the MPEG-2 encoding will happen later – a much preferred option that I recommend you look into. Check the documentation on both pieces of software.

One track, one DVD, no menus

I suggest you record no more than one tune per DVD. Even multiple mixes of a tune on a single DVD are confusing in the darkness of the DVJ booth. One track, one tune, no menus is the ideal. Most new DVD authoring programs, professional or consumer, let you author "autoplay" DVDs that skip the menu when you whack the track into the DVJ-X1. That, too, is preferred, as the last thing you need to worry about is scrolling through a series of menus to get to your track.

Dubbing to DVD

If, like me, you're using a standalone video recorder, your options are more broad and infinitely easier. If your DVD recorder has Firewire input, then running your cable straight from the computer to the recorder might let you do a straight dub to DVD or the hard drive. Might! You'll have to experiment and find out. If it does work, think twice next time you consider updating your software. Updates often fix problems, but they also introduce restrictions that may not have existed in previous versions of the software.

If that doesn't work, connect the Firewire cable from your computer to your camcorder, and use the analog outputs of the camcorder to connect to the inputs of the DVD recorder. What you're doing here is using the camera to convert digital data back to analog – and then the recorder converts it back to digital. No, it's not ideal, but it almost always works, and beats the hell out of rendering a DVD overnight on your computer. You may need to fiddle with the input/output options of the software, camcorder and recorder to make this work. Specifically, enable anything that says something like "play through" or "monitor input".

Finally, if this fails, dub your video back to tape via the Firewire cable, then dub the tape onto the DVD via analog. Check your DVD recorder audio settings to insure you are recording without compression and your levels are perfect (that means no going into the red the way you do in the club). Make the dub, finalize the disc so it can be played on another machine, *et voila*, your first DVD single – done the hard way.

LIVE VJ MIX

Depending on your style, your DVD music video will have several edits – as many as a thousand on a seven-minute track – which eats up significant work time when you're trying to bang out a couple of

videos a day. A quick and intuitive editing solution is to use VJ software to fire your video clips live, and record that mix straight to DVD.

VJ software authors are a bit like bedroom producers – they often make their work for themselves and their friends, given there is no significant market for their product yet. This makes for some of the most interesting tracks around, as well as a lot of crap.

Live VJ mix

The same is true for VJ software, only in this case interesting is not necessarily good. "Interesting" means there is no standard for how VJ software works – and there are dozens of pieces of software. The most common model is VJ-software-as-video-sample-player, as in the more professional packages like *Vidvox Grid Pro, Arkaos, VideoFlux, Resolume, motiondive* or *FlowMotion*. These programs let you collect a group of video clips and fire them from either a computer keyboard or a MIDI keyboard connected to the computer via MIDI or USB.

In general, you'll need analog video out from your computer to record your performance. That's the S-Video connector on your laptop. Some new programs have Firewire output. This won't really work here, even with the Firewire in on your DVD recorder, because you want to record audio simultaneously from the DJ mixer, not the laptop.

Choose your weapon, hook it up

Choose your tune and your software weapon, collect your clips and program a great VJ set. Connect the video output from your computer to the video in on your standalone recorder. Connect the output from your DJ mixer to the audio inputs on your recorder. Again, check both the audio levels and that compression has been turned off on the recorder. Start your tune and rehearse your mix a few times. Pay special attention to the breakdowns and the big moments. Now start over, press record and perform your VJ mix. At the end, you'll have a dynamic, VJ-style music video.

If this description sounds too vague, it is, because no two pieces of VJ software work the same. You'll need to dig into the user manual of your VJ software, and find the details about video formats (codecs), MIDI connections, clip firing and effects – just like you had to read the manual on your first CDJ-1000 – except this is more difficult. I've dropped in some extra information about codecs and the like later, for when you get stuck.

Hard drive + DVD vs. skint

Two points: first, this is where you'll be glad you spent the dosh on the hard drive version of a DVD recorder. Chances are you won't like the first, or even the tenth, take of your VJ mix. Without the hard drive, each take will cost you one wasted piece of DVD media. At a ratio of 15 bad takes for every good one, it'll cost an extra $500 to fill a modest record box. With the hard drive, you can record each take, choose the winner and lose the rest, then make a dub to DVD.

Finally, working this way (the VJ software to recorder method) means a walk into the dark forest of video codecs (compressors/decompressors) and formats. There are far too many codecs. Keeping track of which codec is best for which VJ application is a headache. Programs like *Cleaner* on the Mac or *Video*

Cleaner on the PC will automate all of your format conversions, but they take time and often degrade video quality. You'll also need to determine the size of your file in pixel dimensions. Again, *Cleaner* on the Mac or *Video Cleaner* on the PC can make these conversions, but at a cost. If you have a decent budget, have a look at purpose-built club media machines like Green Hippo's *Hippotizer* line, which simplify the whole process.

LIVE MIX FROM DVD DECKS
DVJ-X1s as production tools

If you want to create your own DVD clubland video singles, chances are you have or are about to buy the finest equipment available for making your own DVDs – a pair of Pioneer DVJ decks, a CD deck, a video mixer and a DJ mixer. While these are obviously great tools for playing out, they are also the best tools for DVJ music video production.

Codecs

Codec stands for compression/decompression. Almost all video files are compressed, otherwise a single feature-length film might take up the space of an entire hard drive. Video players and VJ software decompress the files while playing them back.

It probably won't surprise you that not all codecs work in all software. It also won't surprise you that codecs change as quickly as a DVJ's playlist.

Your best approach to video codecs is to read the manual of your VJ software, find the recommended codec and convert all your video to that format using *Cleaner*.

Here are a few major VJ codecs, along with my guarantee that many will be obsolete by the time you read this:

Sorenson Video – a high quality codec for Apple *Quicktime* video on the Mac.

Cinepak – old and crap, to be avoided.

Indeo Video Interactive (IVI) – a high quality codec for Windows only, but also plays in *Quicktime* for Windows.

MPEG-2 – the codec used for encoding DVDs.

Motion JPEG (M-JPEG) – very high quality cross-platform (Windows and Mac) codec supported by several VJ apps.

Note: for an updated list, try http://www.siggraph.org/education/materials/ HyperGraph/video/codecs/Default.htm or Wikipedia.

Many an audio mash-up and remix are made with nothing more than a basic DJ setup and recorder – one of the most intuitive approaches for the producing DJ. The same principle applies when making a DVJ DVD single. It's not only the best method, it's the most simple method.

Mute DVDs

Start by taking your favorite footage and turning it into short "mute DVDs", that is, DVDs with video but no audio. Break your video down into categories. For example, make a series of mute DVDs with titles like "girl on the beach", "furry bra dancer in the club", "sex in the Mini", etc. They don't have to be long – three to four minutes is more than enough (except for sex in the Mini).

You can create these DVDs using one of the two methods above: the editing method or the live VJ mix. Just skip the parts about importing an audio track. Alternatively, if you're using licensed video already

Furry bra dancer

on mute DVDs, you can skip this part of the process (it will make sense in a moment).

Set up your mute DVDs, self-made or licensed, in a DVD wallet or crate as you would your CDs for a DJ set, only this time you're going to do a VJ set using DVJs.

Choose the tune for your music video, whack it on the decks and get ready to mix the mute DVDs along with the music, using a standard video mixer. You can also add a dedicated effects unit into the video chain, like the Korg Kaoss Entrancer, for more creative expression.

Bosh away

As with the live VJ mix, take the video output from your video mixer and plug it into the video input on your DVD recorder. Do the same from your audio mixer or CD player, this time to the recorder's audio inputs. Again, check your levels and compression. Now press record and bosh away. Don't be afraid to repeat some footage from music video to music video. Over the course of a dancefloor mix, people like to see the same characters or scenes return again and again. Just as music is based on repetition, so too is dance video, and an endless non-repeating stream can turn into information overload.

FINALIZING YOUR DVJ DISC

Many standalone DVD recorders are set up so you can record an episode of *Deadwood*, put the DVD on the shelf and then pull it out again to record another episode on the same disc next week. The hitch is, you can't play that DVD on any other machine unless you "finalize" the disc. Once you've finalized it, you can't record anything else on the disc.

If you do a straight recording of a DVJ track onto a disc – or make a dub from the hard drive onto a disc – you may think you're finished, but once you put it in the DVJ, it won't play.

Be sure to go through the process of finalizing the disc. It's usually a menu-based process, but check your manual.

HOW TO MAKE THAT DVD WORTH WATCHING!

Anyone can play a DJ set, but a quality set is a work of art. The same goes for your DVJ music video.

Structuring a DVJ music video

Just as a great dance track has a song structure, so too does a great DVJ track. For all the intro– verse–chorus– breakdown stuff, go back and have a look at the chapter on song structure. Then apply it to the way you think about the structure of your video.

For example, if you're putting together a DVJ House track, there's usually a long drums-only intro and outro on the audio track to make it easy to mix into another track. You'll want to do the same for the video. Lay down something visually simple to go with that drum beat – one of my favorites is a close-up of smoke against a black background trailing up from the end of a cigarette. This mixes well over almost any video.

When you get to the chorus, you'll want your video to get big – and when you get to the chorus again, use that same video. The chorus repeats: the video repeats!

And don't forget something atmospheric for the

[Eclecctic Method]
The most important new piece of technology for the future of the club scene, and audiovisual performance and production. Hardware that seems like it was invented by God in answer to our prayers. A black chunky beautiful thing which still has so, so much potential for new forms of AV creativity.

breakdown – maybe even the video you used for the intro/outro, but with a bit of effects.

DVJ mute-VJ mixing tricks

Fun, fun, fun, using the features of the DVJ!

- Scratch the video.

- Reverse the video.

- Slow the video down to nearly still for the breakdown.

- Speed the video up as you go into the build, then grab the platter and give it a spin as you explode your way into the chorus.

- Loop your favorite bits and put them on the loop buttons, then switch back and forth between them in time to the beat.

- If you have a video of a dancer, run your finger around the platter on the one of every four beats, speeding her up slightly to emphasize the downbeat.

- Set your camera up on a tripod looking across a quiet road. Shoot until a few cars go by. Once you've got the footage on DVD, set up cue point "A" with just a shot of the road, cue point "B" with a Mercedes driving by and cue point "C" with a jalopy. Press A, then on the one of every measure, press either "B" or "C". Cars go by in time to the beat. You can do this with fish in an aquarium, girls on the beach, or just about anything.

Whatever you do, give it your best mix while recording in the studio so you're always happy to play the track in the club. If you're not completely satisfied, go back and do it again.

And don't forget to have a look at the chapter on video effects.

EFFECTS ARE TO DVJS AS CYMBALS ARE TO DRUMMERS – THEY ARE YOUR GREATEST SOURCE OF EXPRESSION.

The most basic effect is EQ – knobbing up and down the bass, mid and treble to add impact to parts of your tunes. And while EQ on audio is one of the most exciting ways to rock your DVJ set, between the host of audio and video effects boxes out there, EQ only scratches the surface.

Audio effects

The best DJ mixers have send and return options so you can send sound from a channel out to an audio effects box and bring it back in on the same channel. Many of those mixers, like the DJM-800 and the DJM-600, also have an array of effects on the mixer itself. And for lesser mixers, you can always pipe your deck straight through an effects box into the mixer – but you'll need one effects box for each deck.

Clubs are funny about audio effects. They want their DVJs to give the best performance possible, but they don't want you mucking around with their audio setup. I say "Too bad." Get in there and insist your effects are an essential part of your mix. Demand a mixer with send and return. Make it a part of your rider. The worst they can say is no, but if you don't ask, you don't get.

EFFECTS UNITS

These days, there are dozens of standalone effects boxes built for DVJs, and you should use whatever works for you, but my favorites are the Pioneer EFX-500 and EFX-1000. Why? Aside from quality of effects, build and portability, I like them because, like the DJM-600 or DJM-800 mixer, they are the units you're most likely to see installed in a club. If you're an expert at the Alesis AirFX – something I recommend you avoid– then when you rock up into the club and they have an EFX installed, you won't know what to do. Half your effects licks will be useless. And when

[DJ Roonie G]
*DVJing has opened many doors for me. I am breaking
a lot of new ground by mixing not just music videos
but also videos of pop culture fused into my night.
Clips from extremes such as Kill Bill to Dave Chapelle
are thrown into the mix over music that creates a
whole new level of DJ entertainment. The creative
possibilities are endless, since video is involved. For
example I have a video mix of Napolean Dynamite
dancing on top of Survivor's Eye of the Tiger. The
audio and the video – when combined – is taken to a
whole new level.*

you tell the club you brought your own effects, they'll tell you to use the EFX or the effects built into the mixer.

I also like the EFX units because you can reproduce the effects easily. On units with infrared sensors or other gimmicky controllers, you might look cool using them (or not) but you'll have a damned hard time reproducing any single lick from set to set. On the EFX, your effects can also be isolated to work only on certain frequencies – a real breakthrough – and can also be tied to a bpm, which the EFX automatically detects.

Finally, the EFX units aren't gimmicky. To get great sounds, you press buttons and spin platters, not wave your hands in the air looking like a knob.

The EFX units are the foundation tools of digital scratch, house and trance sets alike. I've seen Erick Morillo, James Zabiela and dozens of others toting around the EFX, and to be honest, I do too.

THE SOUNDS

First, all effects are best used in one way: sparingly. Overdo effects and you'll kill the groove deader 'n Skiffle (erm, an Olde English form of Rock 'n' Roll). I usually limit myself to one big effects sweep per tune and a couple of small accents on the four – and then only a few tunes per night. Either that, or I reserve the effects for a big routine three-quarters of the way into my set – and then leave it alone the rest of the night. In either case, the key is the same. Be discerning with your effects and you'll leave them wanting more.

There are several effects that are commonly used in DVJ sets, some of which can be tied to bpms and some of which are best reserved for dialling up on a set of knobs. In all cases, you change the parameters of the effects – the "wetness" of the effect, the "depth" of the effect, etc. I suggest cranking up a tune, pressing buttons and twiddling knobs. Wail away until you find an effect you like, then figure out how you did it and do it again. But do all that in the studio. Keep it tight for the dancefloor. Here's a little info on certain effects to get you started:

Delay

Like most effects here, delay does what it says on the box, box, box .... delay takes a sound and holds it back a bit, then releases it with a repeat. Many effects are just different versions of delay.

Echo

Just like delay, but rather than holding back the sound, it lets you hear it the first time and then repeats it until you want it to stop.

Reverb

An ultra-fast delay. So fast, it runs all the repeats together until it sounds like one big boom in a huge empty room.

Pan

As in panoramic. Panning is moving the sound from one speaker to another, left to right, etc.

Leslie

Leslie as in Hammond B3 Organ with a Leslie speaker. Leslie speakers had a rotating horn on top that swept the sound quickly across the room, kind of like the light on top of a police car.

Phase

Phasing happens when you double up a sound wave and then shift one of them slightly. The sound tends to burble. Think Parliament Funkadelic's *Sir Nose d'Void of Funk*.

Flange

Flanging is the same, but the burble slows down to huge sweeping waves. Flange the sound just right and it practically disappears.

Transform

Like a transform scratch (see digital scratch). Chops the sound rhythmically between silence and full volume, usually in time to the beat.

Filter

Filters only let a certain part of the sound through. Low-pass filters cut out everything but the bass. High-pass filters only let highs through. Etc.

Pitch shift

Think Minnie Mouse.

Distortion/fuzz

If I need to explain this to you, perhaps you should trade in your decks for VIP access to the retirement home.

Video effects

There are two varieties of video effects units – as there are of audio effects, for that matter: plug-ins for software and real-time effects, usually generated by hardware. There are literally hundreds if not thousands of plug-in effects. You'll find them in programs like *Adobe After Effects*, *Apple Final Cut Pro*, *iMovie* and dozens of other programs, and they range from fades and wipes to liquid 3D programs that will map your video across a model of the Atlantic Ocean complete with sunset.

But for DVJs, real-time effects are where it's at. Not that plug-in effects aren't useful when it comes to producing your mute DVDs for use in your studio mixes. It's just that I prefer to use effects live in the club, based on the feel of the night. For this kind of thing, you need a box that gives you real-time effects.

The Edirol V4 is one of those boxes. Although it's a mixer – the video equivalent of the DJM-800 – like the

800 it also has a range of effects. Another solution is something like the Korg Kaoss Pad Entrancer, an effects-only unit with an XY pad for changing the effects parameters. Like the Kaoss Pad, though, the XY interface makes it difficult to reproduce the effects licks you've taught yourself in rehearsal.

These are only two of what will be dozens of units released over the next year or so for the DVJ. Again, find the unit that works best for you – but if you sense that one of these effects boxes will become a club installation standard, buy and learn that one first.

THE VIDEO EFFECTS

Generally, audiences can handle a slew of video effects much easier than audio effects, so you can jump in there and go without killing the groove. Here are a few, but believe me, there are dozens more:

Negative

Like a film negative. Black goes white, white goes

black, red goes light blue (cyan), green goes pinkish (magenta), blue goes yellow, and everything in between.

Monochrome

Takes your color video and makes it look like it was shot in black and white.

Noise

Think 4 am after the television station signs off (do they do that anymore?). Usually, you can dial in how much noise you want to mix across your video image.

Emboss

Emboss makes your video image look like it has been etched into stone or metal. An animated stone tablet!

Binary

Reduces everything to crunchy black and white, with no grays in between.

Mirror

Splits the image down the middle and flips half of it over onto the other side.

Delay

Like audio delay. If someone is running across the screen, a trail of their image will follow them.

Blurring

Makes everything look like it's 4 am in Ibiza.

Stroboscope

Like a strobe light. Flashes the video image at a preset interval.

Feedback

Plug your camera into your TV, then point the lens at the television screen. You'll get a twisting time tunnel. Feedback simulates this effect, throwing an image down the tunnel.

Mosaic

Breaks the image down into solid color tiles. Like what they use to hide people's identity on the evening news. (Squint and you can tell who they are anyway!)

Old film

Makes it look like your video was shot on film, complete with scratches, uneven exposure from frame-to-frame, dust balls and loads of beautiful granularity.

BPM freeze

Freezes the image with the kick drum.

Flip

Flips the image in reverse. Great as a beats-driven effect.

Spin

Spins the image around. You control the speed.

Color reduction

You see millions of colors normally. This reduces the image to a dozen or so.

Tile

Repeats the image in a grid across the screen.

Wave

Looks like your video got the wobblies.

Ripple

Looks like your video is being projected onto water with a little rain.

Time slice

Takes a slice of the image and repeats it across the screen with delay. One of the most beautiful and versatile video effects around. Great for breakdowns.

SANDER KLEINENBERG IS ONE OF THE WORLD'S MOST INFLUENTIAL DJS, REMIXERS AND PRODUCERS, AND IS, QUITE FRANKLY, MY FAVORITE. I'VE HAD THE PRIVILEGE OF PUMPING VISUALS ALONGSIDE SANDER AT GIGS AS DIVERSE AS THE UK'S HOMELANDS AND RUSSIA'S FORTDANCE. HIS "TRANCEGLOBAL AIRWAYS" MIXMAG CD LANDED ON OUR EARS LIKE A BOMB. IN 2001 HE WON THE PRESTIGIOUS MUZIK AWARD FOR "BEST RADIO 1 ESSENTIAL MIX", AND HE IS A STAPLE IN *DJ MAG*'S "TOP 100 DJS POLL". HE MOVED QUICKLY AWAY FROM THE "TRANCE DJ" LABEL, SHOWING HIS ECLECTIC TASTES THROUGH HIS OWN MUSIC PRODUCTION AND MOST RECENT MIX CDS. HIS REMIXES ARE AN ESSENTIAL PART OF THE MOST DISCERNING DJ'S SETS, NO MATTER THE GENRE.

As if all that were not enough, Sander also embraced the AV revolution from the start, and quickly became the planet's most visible DVJ, with audio-visual residencies at four of the world's leading clubs: Pacha (Ibiza), Crobar (NYC), Ministry of Sound

(London) and Panama (Amsterdam). Sander spoke to us from his home in the Netherlands.

Sander Kleinenberg:

In the Netherlands I've been producing parties for more than 10 years now, and visuals have always been a major part of my nights. I work with a VJ who brings the DVJ decks and an Edirol mixer. We also bring two laptops.

DVJ content in the club

Obviously it's not me alone – *Mark Pistoor* is my VJ, my visual guru. Not all tracks that I play out have a video made for them. You have to make your own for tracks you play out more than once. Twenty-five percent of my DJ sets consist of DVDs.

I go through my music, which I still find is the most important part of my show, and then select about five or six tracks a month to shoot. Lyrics are easy to visualize, so often these are tracks with lyrics. Then we think how we feel that the track should be represented.

We have our friends dress up as crazy punks and lip synch the tracks that we've picked out. Or we have words come up for a certain song that can be given impact with text on the screen. We always use the same *typography* and coloring. We work within a fixed idea of what the *color palette* should be. For instance, we've learned that white doesn't work well on the screen in the club. Clubs want to be darker.

Mark, who is super-quick on the machines, edits it all together. He usually starts with one pattern around the beat, like a hand clapping, and then works around that core. We work with the breakdown, the build-up. We'll use bright lights when the rhythm kicks back in.

One of the other techniques we use is to play acapellas and virtual MCs. We might visualize the words, or mimic the acapella, and they become a part of my set. We drop in the audio and visuals as tools for part of my set. We're pioneering and it's a new form, so you have to come up with your own solutions.

You can make video very easily, and it's not an expensive thing. With a laptop, and a simple camera and software filters, you can make something that looks really cool.

I just remixed *Eurythmics*, and I remixed the video as well as the track. I ask a record label if they could provide me with the video. At the moment record companies are not interested in the remix of video, just the track. But it is part of their future – it is coming.

Artistic innovation

An artist is going to be someone who delivers the whole package, who adds video to their product. I had no problems getting with video at all. I had

been looking for a product and process like this for some time. I get paid a lot to play other people's records. If I want to stand out of the crowd and be different, I need to add an element, to be more than a keen record collector. I want to make a Sander Kleinenberg DJ set something you have to see if you want to get the whole thing.

When the DVJ-X1 was launched, I was on the phone five minutes later to get one. There are some techniques that allow you to have a MIDI click from your mixer that you can use with visual software, but it's nowhere close to the power of mixing and creating your own DVJ video. So it was not only an easy step, it was very welcome.

Anyone who wants to do this in a contemporary way, from a DJ point of view, who is not tapping into something that personalizes who you are, you're not seriously thinking of making an impression in the days to come. The DJ being an exclusive selector of music that they have an insight into – those days are over. Looking for that one thing that makes you stand out has become tougher. With white labels, it could be a track that no one has found. But with MP3s and the internet, the exclusivity of vinyl is gone and you have to find it through other means. Vinyl has turned into visual. For me, this is part of who I am. When I release singles, I now release DVDs. The

marriage of the senses is there. Some people in Holland are even exploring scents, to see if this can be a part of entertainment. If we're not all trying to become of the future, it will never happen.

DVJ MASTER

CHOOSING THE RIGHT REMIX IS MORE THAN HALF THE CHALLENGE AND FUN OF DVJING. MAYBE THERE'S A TRACK WITH A VOCAL YOU LOVE, BUT THE MIX JUST DOESN'T WORK FOR YOUR STYLE AND YOUR AUDIENCE, OR YOU LOVE THE BASSLINE OF A TRACK, BUT ABSOLUTELY HATE THE TRANCE SYNTHS. THAT'S WHEN IT'S TIME TO FIND THE RIGHT REMIX. DROPPING IN A ROLLING GROOVE THAT NO ONE RECOGNIZES AND TURNING IT INTO A TRACK EVERYONE KNOWS IS GUARANTEED TO GET A FEW HANDS IN THE AIR.

If you can't find the right remix, don't worry – it's even more fun to make your own. Now that every laptop is just as capable of video production as audio production, the AV remix is becoming as common as your dad's 12" singles.

Ideally, a remix is done with the sanction of the artist and label. This not only gets you paid, it also gives you access to isolated parts of the original track –

the vocal with no instrumentation, the bassline without drums, etc. But the mash-up (UK) or bootleg (US) is just as common, and has the advantage of giving you total creative freedom. The disadvantage, of course, is that if you release it, you might get sued.

The tools

The tools for remixing are the same as those for production, and you'll find a basic outline in the chapter "Choosing the Right Software". You'll find each piece of software equally useful, but some are more equal than others.

AUDIO

Ableton Live is the ultimate tool for remixing dance tracks. Not only is it loop based, thus specifically built to accommodate dance producers' preference for repetition and samples, it also enables VST plug-ins. VST means you can bring in the staple of most dance-music sounds – a variety of effects. It can take hours of lyrical composition and arrangement to cover 64 bars of a pop track – in dance, it sometimes takes nothing more than a decent filter sweep to get people's hands in the air for the build-up. That's where the plug-ins show their real value.

VIDEO

I like a combination of tools for creating video remixes. Obviously, you're going to need an editor like *Final Cut Pro, iMovie* or *Premiere*, but I also think a VJ tool is indispensable. Bits of the original music video can be broken down into individual shots, stored as video clips, and then triggered from *Grid Pro* or some other VJ program using a keyboard.

It's more than possible to reproduce this effect using nothing other than a traditional video editor – just as it's possible to reproduce scratch effects through laborious editing in *Final Cut Pro*. But why do that when you could just scratch video on the DVJ and record it? The same goes for rapid-fire MTV style editing. Why slog over placing each edit against the beat in a timeline, when you could just fire the clips from your computer keyboard in a VJ app and record the performance?

Even if you only use these approaches to get a rough edit, and then go back and tweak in an editor, you'll find yourself with a far more intuitive music video re-edit by using tools made for the marriage of music and video.

The essence

A great dance remix is usually a process of reduction. Despite what your parents might have told you, most pop songs are complex, particularly in R'n'B. Arrangements can be thick and lush with unusual rhythm and meter. Melodies are often far more complex than on the typical dance track.

What's worse, the original producer of the tune might actually be a hero of the DVJ who now has the weighty task of adding something new to what they already think is a brilliant track. This is the trap.

The way out is to remember this: a remixer's job is almost always about reduction.

A remixer's first questions are: what is the essence of the tune? What is the hook? What drives the track and makes it unique? Once you know, sample that phrase straightaway.

The second priority: how do I make it groove?

Less is more

The Infusion remix of Kate Bush's *Runnin' Up That Hill* is a great example. The original Bush track is lush, with a complex and lengthy lyric.

After laying down a basic groove, Infusion bring in a new bassline – a single note that throbs on the downbeat.

The lyric then comes in as a build-up – "Only could, be runnin' up that hill." And that's it, repeated over and over! The outro of the original Kate Bush track becomes the principal lyric for the entire remix.

The bassline then changes to take the tune through its chord changes, while the groove continues in the background.

At the breakdown, the lyric returns – "Come on baby, come on darlin', let me steal this moment from you now, come on baby, come on darlin', let me steal this moment from you now, now, now, now, now, now …"

Lyrically, Infusion leave it at that. The rest is breakdown, build-up, chorus and groove. Possibly one of the best remixes ever recorded. Genius in its simplicity. Guaranteed to put their hands in the air in its day. And it's 9:27.

Another slightly more complex track is Sander

Kleinenberg's Smokin' Dub of Annie Lennox's *Wonderful.*

Here again, the groove is laid down first, becoming more complex through the intro. Sander often uses vocals as nothing more than sounds – in this case, after giving the track an urgent bassline, he brings in Lennox: "Does it feel feel feel feel, does it does it does it." A single line, echoed from different words, emphasizing its simplicity. "Does it feel hard." This is the chorus, and Sander renders it in an instrumental way.

After a repeat, we hear a lead-in to the next choral variation, "Smokin' like some crazy fire," used more as a structural accent than a lyric.

At the breakdown: "But I feel wonderful."

Just to reiterate the point, here are the words of both of these lyrically complex tunes once they've been stripped down and remixed:

Kate Bush – *Runnin' Up That Hill* (Infusion remix)
Only could, be runnin' up that hill
Come on baby, come on darlin'
Let me steal this moment from you now

Annie Lennox – *Wonderful* (Sander Kleinenberg Smokin' Dub)
Does it feel hard
Smokin' like some crazy fire
But I feel wonderful

In both cases, simplicity is the essence of the lyrical approach, and in remixes the same often applies to both the instrumental and visual structure of the remix. Listen to a few of your favorite remixes, discover their simplicity and then apply it to your own remix of your favorite track.

THE PIONEER DVJ-X1 AND DVJ-1000 ARE THE PRINCIPAL TOOLS OF THE MOMENT FOR DVJS, AV ACTS AND JOURNEYMAN DJS WANTING TO GET INTO VISUALS. ON THE SURFACE, IT'S A STRAIGHTFORWARD MACHINE WITH THE MOST MINOR LEARNING CURVE FOR DVJING, HOWEVER, FOR THOSE WHO LIKE TO GET UNDER THE SKIN OF THEIR MACHINES, THE DVJ HAS MUCH TO OFFER. FROM CUSTOM TRAVELING SOLUTIONS FOR TOURING DVJS, TO DEEP LEVEL MENUING SYSTEMS, AS WELL AS NAVIGATION OF DISKS CONTAINING HUNDREDS OF MP3S, I'LL WALK YOU THROUGH SOME OF THE MOST IMPORTANT FEATURES AND EXTRAS. KEEP YOUR MANUAL HANDY.

THE DVJ-X1

A quick look at the DVJ-X1, this video walks you through a few of the fundamental controls of this demon digital deck:

- Comparisons to the CDJ-1000
- The platter
- Bumping and dragging tracks
- Adjusting tempo
- Master tempo
- Cueing
- Recording and playing back from hot cues A, B and C
- Starting up and winding down

- Cueing and looping
- Relooping
- Storing loops in A, B and C
- Re-cueing
- Firing from cue
- Saving cue points to memory cards
- Playing with pitch using the tempo buttons

For an in-depth look at the DVJ-X1, see the chapter "Getting the most from the DVJ-X1".

The touring DVJ
MONITORS AND CASING

Depending on the distributors in your country, different versions of the Pioneer DVJ case have been designed specifically to protect the precious DVJs. In some countries like the UK, a monitor system is included with the case. In others, you'll find different features.

These all have advantages and disadvantages, and with the internet you can just as easily order the Japanese version from Portland as you can the UK version from São Paulo. Take your pick, or build your own based on a standard turntable case. Either way here are some things to look out for:

- Monitors. Building monitors into a case is an exercise in trade-offs. On the one hand, there are several reasons to avoid cases with monitors: most mixers let you preview what's on a channel before putting it into program (thus you only need one monitor for a whole DVJ rig); adding monitors to your case adds significantly to the weight; and chances are you know what's on your DVDs because you'll have made them yourself. On the other hand, accessing cue and loop points is best done through the on-screen menus which are only available through the DVJ's preview monitor output, although it is manageable without the monitor. Also, you'll need access to the DVJ's menu system if you're going to navigate MP3 discs. The choice is yours.

- Build quality. A light metal turntable case may perfectly protect your DVJ from shocks and bumps, but if a rod can be shoved through the thin protection on the side of the case, your DVJs will likely suffer damage. Neither airport security nor airline baggage handlers give a toss about your kit. If you're the only person handling your cases, a lightweight turntable case is fine. If you're going to place your decks in someone else's hands, then I suggest either a custom-built case, or the superb Peli 1550. Doctors, scientists, engineers, photographers and videographers regularly use Peli (or Pelican in the US) cases. They're waterproof, submersible, strong as steel despite their light weight, and even maintain air pressure through an airlock system. The 1550 accommodates the DVJ perfectly. I've flown the world with a pair, and so far so good (except for a very recent run-in, when US Homeland Security decided to destroy my decks outright! Buy insurance). Just be sure to place foam above and below the jog wheel, so nothing rubs against it in transit

- Protection. There are three key issues in protecting your DVJ in its case:

  1. Nothing can touch the platter, and preferably none of the buttons
  2. The DVJ will neither move nor be shocked by a bump
  3. Nothing can penetrate the case.

- Weight. Airlines have different weight limits for different journeys. Internal European flights hover between 20 and 24 kg. Go over that weight and you'll pay as much as £40 per kg. Two DVJ-X1s in lightweight cases clock in at about 25 kg. Go for the lightest case you can find that is still capable of protecting your gear, or see your profits dribble away.

- Cost. If, like me, you're a motorcyclist and you're trying to decide how much to spend on a helmet, here's a good question – how much is your face worth to you? How about your skull? Right, that'll be a full-face Bell helmet for me then. The same applies to cases for your DVJs.

PAL, NTSC AND REGIONS

In an ideal universe, you would be able to take your discs anywhere on the planet and play them back in any DVJ-X1 system. Unfortunately, it's not an ideal world, it's a world split into PAL and NTSC, as well as DVD regions.

PAL and NTSC

As we saw earlier, there are different video standards for different parts of the world. Most DVD players will play only one type of video – PAL or NTSC. The DVJ-1000 goes a step further, covering from one format to other on the fly. The DVJ-X1 plays both, but with some limitations. If you put a PAL DVD into an NTSC machine, the DVD will play, but the machine will output PAL. And vice versa for NTSC in a PAL machine: the DVJ-X1 will output NTSC.

This is fine in certain circumstances. If, for instance, you are mixing video using a V4 (which accommodates both PAL and NTSC), and you are feeding an up-to-date projector (almost all of which accommodate both standards), then your video will play back as you expect it to. But only if you don't mix discs. Why? First, the V4 has to be powered down and back up again to switch systems. Second, even if this were not the case, the projector would lose sync in the transition from one format to another.

Also, if there is a video distribution amplifier between the V4 and the projector(s), you're screwed.

This is not the end of your problems. If you're playing commercial discs, there are DVD regions to contend with.

DVD regions

The good folks who don't want you to pirate DVDs have carved the globe up into six DVD regions. Discs from one region generally won't work in players from another region. Why? The given reason is that distribution companies need to control the release of their films on DVD. Most Hollywood DVDs hit the American market before they arrive anywhere else. Should someone from China fly to a Blockbuster in Nevada and take home a newly released DVD, they

DVD regions

0 No region coding

1 United States of America, Canada

2 Europe, including France, Greece, Turkey, Egypt, Arabia, Japan and South Africa

3 Korea, Thailand, Vietnam, Borneo and Indonesia

4 Australia and New Zealand, Mexico, the Caribbean and South America

5 India, Africa, Russia and former USSR countries

6 People's Republic of China

7 Unused

8 Airlines/cruise ships

9 Expansion (often used as region free)

could play it in to their home country before the distributors officially release the film there. Horror!

There are workarounds. Many players have "region hacks" – code sequences that can be tapped into the DVD player via the remote control to make it region free (like tapping a game hack into Playstation). You'll find these on the internet by Googling your brand and model. Also, when you buy a normal DVD player, particularly outside the US, the salesman is often happy to tap that code in for a few quid extra. Unfortunately, the DVJ is bound by regions, just like a standard DVD player. And there are no region hacks – at all!

What does all of this mean?

The PAL/NTSC issue makes it difficult to play tracks on another continent. Even if you produce your own tracks, they will be authored in one system or the other. You could carry your own decks and mixer (as I often do), but there is no guarantee your kit will match the projection system. I've learned by experience that even a thorough recce cannot prevent gig-time disaster. Unless they have an intercontinental attitude, it's unlikely your on-the-ground club technician will possess intercontinental technical knowledge. To deal with this I often carry a selection of CDs.

As for regions, if you're playing nothing but off-the-shelf tracks, and you bought them all at home in Manchester, you won't be able to play your DVDs in a machine in LA. And vice versa. You'll need to play DVDs you've made. Which doesn't solve the PAL/NTSC issue.

What to do? For this, I have few answers. Making your own DVDs insures they are region free. And a thorough recce or the purchase of DVJ-1000s helps with the PAL/NTSC issue. Just make sure you press for hard and fast answers regarding the club's system, video distribution amps, etc. And say a little prayer before you leave home.

MENUS

The Pioneer DVJ is covered in buttons and knobs. Fairly confusing at first, but by now you'll have got past the willies and be well into your kit. Underneath all of that visible technology are a huge range of features available through the on-screen menu system, and they're worth having a look at. Just like the memory cards, you're missing out if you don't at least give them a try.

You can only see the menus through the preview out video connection. If you haven't been using this – and I suspect you have if you've delved into saving cue points and loops – you should hook it up and have a go now.

Two notes:

- You navigate around the menus using the big round button with directional arrows, in tandem with the enter button. It works

just like the controls on your home DVD player.

- Most of this information is in the manual, and, ultimately, that should be your guide. I'll just take you through some of the more interesting points.

MP3 menus

The Pioneer DVJs play MP3s from discs. This is a boon for anyone who wants to rock up into the club with two discs in their pocket – and still have more than enough tracks to play for several hours.

However, negotiating the disc with nothing more than the standard LCD screen will be nearly impossible. This is one case where you'll definitely want to use the video preview output of the DVJ.

Using the preview, you'll see that you can navigate an MP3 disc CD using the rocker control, much like you'd expect to find on a DVD remote. Pressing up and down moves you between MP3 folders, and pressing right takes you into whichever folder is highlighted. Pressing left takes you back to folder selection.

Keep a couple of things in mind when you're dealing with MP3s. First, you are limited to 99 folders. And each folder is limited to 999 files. So, whatever you do, don't write more than 98,901 MP3 files. I'm kidding, of course. But you do want to keep in mind the structure of your disc. Think carefully about how little time you'll have in the club to find the track you're looking for, and structure your disc accordingly. Also, consider making two copies of each disc. If there are two tracks you want to play back to back, but they're both on the same disc, you're going to be up the proverbial without a paddle. And it's always good to have a back-up.

Wave preview

In the LCD feedback menu on your DVJ, there are some limitations. When you're playing a CD, the menu will

show you a rough waveform of your entire track, plus where you are in that track when you're playing it back. If you're playing a DVD, however, the waveform in the LCD will only show you the wave up to the point where you are at the moment. The latter isn't all that useful. The waveform is used much in the same way you might look at the position of the needle compared to the patterns of the grooves on vinyl. By glancing at the vinyl, you can get a rough estimate of how soon the breakdown, or the verse, is coming. The same is true of the waveform – unless you're playing a DVD.

There is, however, a workaround. By using the video preview menu, you will see the complete waveform, as well as your position within it, whether you're playing a DVD, CD or MP3.

Setup

Setup gives you access to a load of features, most of which will be obvious (like languages, etc.), but there are a few worth a further look.

Dolby Digital audio

This will be useful for you only if your tracks are in Dolby Digital audio, a legacy from the analog days. If they are get in there and match the features between your discs and your DVJ.

Surround sound

Just a note here that you won't be able to use Surround sound if you're planning on DJing with your DVJs. Surround sound works in normal mode only, and is principally aimed at playing back movies.

Legato PRO

You can fool around with this in your studio and see which setting you like most (I like Effect 3 – lots of bass!), but keep one thing in mind: at the club, this feature will be turned off, and they won't want you turning it on. I prefer to leave it off in the studio.

Hi-bit audio

Your DVJ is a true 24-bit machine and can upgrade the bit rate of your tunes to a higher level. Take advantage of it if you can. The DVJ is capable of 24-bit / 96k output. What does that mean, and how does it affect your DVD's sound?

24-bit / 96 kHz means, roughly, very high resolution output of waveforms up to 48 kHz. That's way beyond the range of human hearing. But few DVDs or CDs are recorded at that resolution. Your DVJ can certainly output at a higher resolution than the format of your disc, but the most you can expect is a little more warmth.

No club on the planet, at this writing, is capable of reproducing 24-bit / 96 kHz sound. But your DVJ is. And so are plenty of mixers. Amps and the like are just around the corner.

Thus, if you're creating your own tracks or remixes,

consider changing the bit rate and frequency range at which you record. It's only a matter of time before this new sonic capability becomes the standard.

Aspect ratios

I spoke about aspect ratios earlier. The main thing to remember here is that if you've shot your work in 4:3, then set your decks to 4:3 and project it in 4:3. Same for 16:9.

What happens when you get to the club and your discs are in a different aspect ratio than the club's system? Get into the menus and reset things.

If the club's system is 4:3 and your discs are 16:9, then set the DVJ to 4:3 (Letter Box). This will place a black band across the top and bottom of the screen, like you've seen on your home TV.

If the club's system is 16:9 and your discs are 4:3, then set the DVJ to 16:9 (Wide). Black bands will appear on the left and right of your image.

If the promoters or technicians won't let you make the adjustment, or won't make it for you, in general you shouldn't worry about it. Unless this is a crucial gig. In that case, demand perfection. Generally, though, things aren't so critical in a club environment as in the living room, and the surest way to not get booked again is to throw a DVJ tantrum over something only you care about. You'll have to balance these factors and make your own decision.

Progressive

You'll rarely encounter a situation where you can use Progressive output. Why? Because very few video mixers will handle Progressive component video – at least none a normal or even super-club can afford. And if you're playing Wembley with a full Progressive video projection system, then I suspect you're probably not having to do your own menu settings.

Screen saver

Turn it off, unless you want an ugly pink Pioneer logo on a white backdrop in the middle of your set.

Video quality adjustments

Don't! Just don't!

A FEW COOL FEATURES

Back to the interface. Here are a few cool features you'll want to get to know:

Auto Cue

When you put your DVD or CD in, provided there are no menus on your DVD, this will automatically cue your tune to the first bit of the track. This can save you loads of time if you want to start the track from the beginning. But be careful – often the deck will hear  "ambient" noise on the track that it thinks is the beginning. Double-check before you drop it in the mix.

Hyperjog mode

This doubles the response of the scratch wheel. Great for fast forwarding through tracks, but not so great for subtle scratching. Play around and get to know it.

Touch/Brake and Release/Start

 This is a vinyl-emulation feature that can be used to great effect in the club. On a standard pair of 1210s, you can turn the power off on the turntable while it's running, and the deck will slowly wind down, outputting audio all the time. Turn the DVJ knobs all the way clockwise, and you'll get a slooooooow start and stop. Full counter-clockwise, and the starts and stops are instantaneous.

Direction FWD/REV

Does what it says on the box. A great effect, used sparingly. Keep in mind, though, it will throw off the phrasing of your track.

Emergency Loop

At the touch of a button, you can grab four beats in a near-perfect seamless loop. This is very effective in case your track is about to run out and you need to extend the outro for a few bars. It's also a good way of grabbing a 4/4 measure anywhere in the tune for a quick loop.

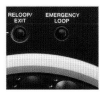 In use, I've found that it tends to grab slightly more than four beats, so you'll need to adjust the out point with the jog wheel or the tempo

slider if you're in the mix. It's a subtle adjustment – you'll only get completely out of sync after three loops – but it's enough that you need to know about it.

By pressing the Emergency Loop button for one full second, the entire DVD chapter will repeat. This is useful if you have a visual (only) on the DVD that you want to loop. For example, I usually make up a DVD that has several logos on it – one for me, one for the club, one for the other DJs – and each logo is animated and recorded on its own chapter. Once I've selected the chapter I want to play, I then put the DVJ into chapter repeat. That can turn a three-second animation into an infinite loop capable of playing all night. A very useful feature.

Text/Wave

Text lets you see the CD or DVD text of your track. This is useful if you've forgotten what track you have cued up – believe me, it happens.

Wave is far more useful, and I suggest you use that as the default position. Wave shows a rough waveform in the display, giving you a visual guide to the peaks, builds and breakdowns of your track. Very important when you're doing structural mixing.

Master Tempo

 Any track rocked up to 170% of its tempo sounds like Minnie Mouse – fabulous for Happy Hardcore and Gabba, not so hot for most other applications. Press Master Tempo and the pitch of your record will stay the same, even while you're adjusting the tempo.

Sometimes when you're making subtle adjustments to the speed of a tune in order to get two in sync, it's more useful to turn off Master Tempo while you make the adjustment, and then turn it back on before you bring the track into the mix. This is very dependent on one thing – how well you know your tune. Do whatever works for you, but I suggest you know your tunes.

DISPLAY

The display is your information feedback zone. Nearly everything you'll need to know about your track, aside from how it sounds, is in the display – and it's well worth learning.

Most of the features are documented in your manual. Here are a few that I use every gig:

Auto cue

Tells you whether you're in auto cue or not.

Remaining time

Gives you the choice of seeing either how far you are into a track or how much time you have left.

Tempo

The tempo of your tune as a percentage of the

original recorded tempo. The more broad your tempo range setting, the less detailed the information.

Tempo Control range

6%, 10%, etc.

Master Tempo

Tells you whether Master Tempo is on or not. This info is duplicated on the button itself.

BPM

Tells you the bpm of the track. It's not perfect, but is a good general guide.

M. Cue

This tells you whether memory points have been set up for your track and where they are, relatively,

within the track. That applies to points stored on the memory card as well.

M. Loop

As above, but with loops.

The DVJ-1000 jog wheel

GOING ALL THE WAY

SASHA AND DIGWEED REVOLUTIONIZED THE IDEA OF THE CD WITH THE 1994 RELEASE OF *RENAISSANCE*, A STORMING TRANCE MIX THAT DEFINED THE PAIR AS THE THEN-LEADERS OF THE TRANCE GENERATION. *RENAISSANCE* TURNED A BOX FULL OF DISPARATE 12" SINGLES INTO A JOURNEY LED BY S&D – A TRUE PERFORMANCE TO MATCH THEIR LEGENDARY MINISTRY LIVE DJ SHOWS.

The DVJ has created an entirely new genre of mix disc – the DVD mix. While some early forays were made into mix DVDs by DJ/VJ teams, nothing beats the DVJ mix, where both the music and video flow from the creative vision of a single artist or group. With DVD standalone recorders, making a DVJ DVD mix can be as simple as rehearsing your mix and recording it. Or you can go all the way and combine your studio tools to create a legendary audio-video epic.

What works in the club doesn't work at home

The first thing to keep in mind when you're working on your DVD mix is that the club isn't the living room. In the club, you're mixing for a group of people who are up for a storming party. They want to put their hands in the air, shake their ass and show everyone what they've got.

The audience for your DVD mix might be a mid-20s bloke knocking back a brew on the sofa in his underwear. He may love the same music your club audience loves – hell, he's probably part of your normal club audience – but on the couch, he's gonna want a different experience.

Whether you're in the club or in the living room, a DVJ performance is about shining a light in different places at different times. In the club, you might point your flashlight at the tune for a while, but then flash it on the audience by dropping them into a long rolling groove. Next you might point the light at yourself with effects tweaks and a bit of DVJ flair, and then throw it back on the audience again, or on the screens.

In the living room, you can shine the light anywhere you want, as long as it's either the music, the video or yourself. What's missing is the audience. And the lights and the bar and the babes and and and ...

How the video differs

- Quick-cut flashy editing doesn't work as well in the living room as in the club. You have to bring the intensity levels down and give the video more space to breathe.

- While there's less room for repetition over short periods of time, there's infinitely more room for using repetition across the mix DVD to experiment with loose narrative.

- Photosensitive epilepsy is a REAL issue in the living room. Read that warning box again.

How the mix differs

- You have to start the mix. In the club, you can mix into the warm-up DVJ's last track. Or start with a repetitive beat to get the audience grooving. On the mix DVD, you need to open the mix. Try a vocal sample from a movie (don't get sued), or a track that starts with a vocal. Create a scratch intro routine. Anything, but make sure you OPEN the mix.

- Long, rolling groove sections are fun in the club, but get tedious on a mix DVD after about 64 bars. You'll want to switch it up a bit – with effects, dropped-in samples or layered tracks.

- Long, deep drops and breakdowns don't sustain as well in the living room as in the club. Putting your hands in the air alone in front of the fireplace just doesn't give quite the same thrill.

- There is absolutely no room for the long intros and outros of dance tracks on a mix DVD. They're energy killers, and you want your mix DVD tight. In the club, it's sometimes a good idea to give the dancefloor a little breather. No one needs a breather from sitting on their ass watching tunes.

Choosing the tunes

Your main focus here should be the flow of the DVD. What journey are you taking your audience on? Remember, you're no longer bound by what's making them dance. You're driving the car on this one. Where are you starting? What's the destination? What do you want to show them on the way? Is this a happy journey or a serious trip? Are you out for a laugh (humor works much better in the living room than on the dancefloor)?

And not to sound like your mom, but when you're whipping up your set list, get all your tunes together in one place. If you're in the basement mixing, and that perfect next track is upstairs in the living room, believe me, you're a lot less likely to go up and fetch it. It'll never make the cut.

Custom mash-ups and mixes

You have all the time in the world to construct your audience's hour-or-so journey. I'm sure you can find time to create at least one custom mash-up or layered mix.

Recording

There are several approaches to recording your mix. Some might choose to drag their DVJ music videos into an editor and manipulate them into a mix there. While there's nothing wrong with this approach, it produces a lot of dry, emotionless mixes.

Far more intuitive is the proper DVJ mix, recorded and tweaked.

Recording a DVJ mix can be done in the same way you made your DVJ DVDs – by hooking up your audio and video mixer to a hard drive recorder or a mini-DV recorder, and wailing away at your decks. Once you've created the basics, you can then take your mix into an audio/video editor like *Final Cut Pro* for a bit of tweaking. If you've recorded onto DVD or hard drive, dump the data via Firewire onto mini-DV so it's easier to then get the video into your computer. Reverse the process after you've edited. Just make sure you're recording PCM audio on your rough mix, otherwise you'll end up with too much compression and level adjustment, killing the groove.

Mastering and tweaking

You have a chance in the editor to add some DVJ flair. Drop in that extra effect that you would normally like to put in the mix, if only you'd been born with three hands. Lay down an audio-video accent on the start of a few phrases you would like to see emphasized. And clean up any mixes you might have duffed (and if

you duffed one too many, go back and re-record the mix now, before you invest a lot of time tweaking, only to throw out this version later).

This is also your opportunity to master your mix. On the video side, set the contrast and brightness just the way you like it. On the audio, get your levels even, remove any DC offset, compress the audio (not too much!) and snip out any superfluous bits.

Watermarking

This is critical. Club promoters suck. Not only do they use DJ's mix CDs as drinks coasters, they use DVJ's mix visuals as free club styling. You may believe in file sharing when it's someone else's work, but the first time you walk into a club and see the video from your demo playing behind some sound-only DJ, you're gonna rethink your sense of generosity.

Create a logo and stripe it across your demo mix DVD. Put it in the corner MTV style, or stripe it across the bottom like it's a subtitle – I don't care, just do it. If it's a demo DVD mix, then be sure to include your contact details in your watermarking.

If your mix DVD is for the public, then watermarking won't really do. Try dropping an animated logo in for a couple of seconds every minute or so. That way it can't be played out in a club without either giving you free promotion, or without someone having done one hell of a long editing job.

Even if you want to give your work away, at least make sure you get credit for it. Protect yourself. No one else will.

The final DVD – menus, menus, menus

The cheesy menus on your standalone DVD recorder are fine for creating your own DVJ DVDs. But for something you want to send out to clubs or to the public, you'll need to bring your video mix into an editor, export it as an MPEG-2 file and import it into a DVD authoring program.

iLife is fine for this kind of thing. I find their pre-packaged DVD menus, if not as ugly, then at least as cheesy as those on hardware recorders – but you can also design your own. Play around and have a creative blast with this. And for demos, be sure to include your contact details on every menu page.

Watermarking

GOING ALL THE WAY

Networking

Networking is the number one way to get gigs. That means lots of clubbing. Get to the club, check your coat and keep your demo DVDs in your back pocket. Hand them over the DJ booth, hand them to the bouncer, hand them to the bar staff – and hand them to the guy who looks like he's in charge.

Get in the VIP if you can, and talk it up.

Here's the main thing to remember – never slag anyone off. The bartender might be the man who books the DVJs. The washroom attendant might be the owner's best friend.

And please, never, ever say anything as misguided as, "This DVJ sucks, dude. I could do a much better job than this loser." You might be talking to her husband.

Agents

Sad, but true. If you're gonna go all the way, you're gonna have to get an agent. Most young (and even many experienced) DVJs think "If I could only get an agent, then I'd start to get bookings."

This couldn't be further from the truth. There are two situations in which an agent might become interested in handling you:

• You are already generating a great deal of income, press and interest, so they know you'll make them lots of money

• You are blindingly, unbelievably talented, have something no one else has, and it's obvious to everyone who sees you play that you are a good investment

Sometimes agents bet wrong. They think someone will make them money, and then realize they won't.

Agents who are great businesspeople will drop said DVJ like a hot potato. Agents who are loyal will stick by you, but not call you very often. A truly great agent will figure out how to build your career, despite what they overlooked.

Find a great agent who will sign you and who you trust, and stick with them. They're rare, and a real prize. And if you're fortunate, at some point, they'll become one of your best friends, just as mine has.

Competitions

DVJ competitions are a good way to get your name out there. Both Eddie Halliwell and Yousef got their breaks winning DJ competitions. Winning will usually get you a bit of kit, put you in front of a few promoters and give you some legitimacy. In other words, you get a break.

You won't get signed to a leading agency, get a headline gig or get made. Those bits are up to you.

Mailing out your mixes

James Zabiela and quite a few others swear by mailing out DVJ mixes to promoters and anyone else with an address. This can work, but it has to be followed up with phone calls, additional mixes and club visits. Just as your computer programmer nerd mate is never gonna make it with EA Sports unless he gets out there and meets some of the right people, you're never gonna conquor clubland unless you get out of the bedroom.

Trade-offs (VJ / DJ + warm-up = gigs)

Here's a trade-off that works, but it's only gonna work for a while. Lots of the old DJs are traveling around without visuals. They don't know how to make them and they can't be bothered to pay for them. But punters are now demanding visuals and the old geezers are in a panic.

So what's the deal? Offer to mix the DJ's visuals live for free. But he has to agree to insist to the club that you warm up for him as a DVJ.

This is a win–win situation. The club has to hire a warm-up DVJ anyway, so why not you? Everybody gets paid (although you won't get paid much) and no one pays anything extra. As good as it sounds, it won't work every time, but when it does, it will be worth it.

And, erm, don't call him an old geezer to his face.

Playing for free

I'll tell you not to do this, you'll go and do it anyway. Just don't do it too often.

Your website

This is your number one marketing tool. A great website will have your mixes, your pictures, your news, your tours, your fans, your moms and your dog on it. It'll save you a fortune in mail-outs. Make sure it's as professional as your mixes. If you don't know how to build one, then make a trade with someone who does. Play their party in exchange for their web-building skills.

Just make sure you get a tight site, that you get it right, and that it's easy to update.

Resources

There are plenty of helpful resources out there. In particular, the Music Industry Manual (MIM), better known as the bible. It's been around for over a decade and has its roots in the dance music scene. It lists over 100,000 contacts for the music industry, including promoters, clubs and DJ agencies.
It costs a hefty £65, but has recently gone online, allowing the subscriber to email mixes and clips to all its contacts. 100,000 contacts in one blow! Pretty nifty. Check out www.mim.dj.

GOING ALL THE WAY

First things first

What you've got in mind is to create an amazing DVJ mix, get it up online, and achieve fame and fortune overnight. Should be easy, no? Well, no.

Entire volumes have been written about how to create streaming video to place online, how to embed it on your web page, how to embed that in your website, and then how to promote your website so you manage something better than a Google whack when you're sitting at home Google whacking yourself. I'd hate to kill more trees over that one. I do suggest you have a look at *iLife*, which automatically compresses and posts video to your website. It's a gem!

What I will also tell you is this – getting that mix up on your website is a GREAT idea! But you need to consider some of the same issues that you would with building your own website; namely, do you know how to build a website? If you don't, you need to call in a pro. Same goes for setting up a streaming mix – if you don't know how, call in someone who does. Or use *iLife*. You'll be better off spending your time perfecting your DVJ skills.

That said, here are a few tips ...

Recording your mix

To get your DVJ mix up online, you're going to need to record your DVJ mix in the first place. I use a fairly straightforward method for this one – the same method I use to create my DVJ DVDs. Hook your DVJ decks up to your DJ and VJ mixers, hook their output up to a DVD recorder, and start mixing. Refer back to the "Going All The Way – Making A Mix DVD" chapter for details.

Next, I suggest you run this mix out to Mini-DV tape. This is the easiest transfer format for anyone building a website. Your web builder can recapture the footage and make the digital conversion themselves – tweaking in a way they prefer. Again, if you don't know how, leave it to a pro.

Technology

And here, I mean end user technology. No matter how good your web builder bod, you're gonna hate the way your video looks online. Don't panic! The low quality of online video delivery is a given, and end users don't expect any better. That said, you DO want to insure that as many people as possible are able to access your mix.

That means you want to use the prevailing online video delivery system – something that works in Windows Media Player and Apple *Quicktime* (if you wanna check the latest formats, just see what those porn portals are pumping – at the moment it's MPEG in a .mov file). You also will need to invest in a decent web hosting service – not the kind you get for free. Buy your own domain. Make sure the host company has a streaming server. Get a contract to allow at least 5 gig of web stream each month.

[Dan Tait]
DVJing is about enhancing the dancers' experience by affecting another of their senses. My DVJ tracks are rhythmic and groove led, just like the music I play. I try to keep things simple and reflect the mood of the music by highlighting poignant elements in the track. I love juxtaposing incongruous imagery and text to play with people's minds. For me, visuals don't compete with music – they are just another language.

Legalities

If you don't own the rights to the tunes you're streaming, you need permission from the record company to use it in your streaming mix. Full stop.

Here are a few things to help you with this incredibly thorny issue:

- Try using unusual tunes from unsigned artists. You'll find their releases flagged up in the chatrooms of peer-to-peer file-sharing networks like Soulseek. Write the artist an email, get their permission and mix away.

- Have a trawl through creativecommons.org. There are a gazillion tunez available for download and re-use under a variety of licensing schemes.

- If the work is on a small label, write the label, tell them exactly what you want to do and ask permission. It's a long shot, but it works often, particularly if your project is more artistic and less commercial.

- Write and record your own tunes.

- Take a risk ...

On this last note – I can't really recommend it. Sure, you might get away with it as long as no one's looking at your website, but then, that's not the goal. The moment you start getting major traffic, you'll start getting major lawsuits. And it's worth keeping in mind that an Italian DJ was recently fined over €1,000,000 for playing illegal MP3s. Don't do it.

Traditional scratch technique

Old skool vinylista scratch technique is a mystery to most non-DJs. What the punters always notice the DJ doing is wicky-wicking back and forth on the 1210s and, in their minds, that's the essence of scratch.

What they miss is what the other hand is doing – gettin' busy on the crossfader.

Almost all old skool scratch is based on a combination of movement between one hand on the vinyl and one hand on the crossfader. The vinyl hand creates the basic sound, and the crossfader hand is used to cut that sound up, creating a volume envelope around the scrrrrraaaaatccchhh of the left hand.

A quick look at Transformer technique makes it more clear. Transformer technique was invented by DJ Cash Money, based on some cross-cutting technique he had seen performed by DJ Spinbad in Philly.

Spinbad would use the crossfader to cut short bits of one piece of vinyl into another. Cash Money evolved the technique by combining the cutting with scratching the vinyl – producing a sound he thought was similar to that of the Transformer cartoon robots.

Here's how you can try it out: find a "white noise" sound on a CD. "White noise" is the sound you hear when your television isn't tuned to any channel – it's static. Load it up in your DVJ and create a loop around it. Now set your crossfader to the "sharpest" setting – the one with the least curve on it. Take one

Digital scratch

Here I show you how to use the EFX units to develop a straightforward digital scratch technique. I focus on the Transformer effect, synchronization with the track you're scratching over, and scratching on the DVJ-X1.

hand and grab the jog wheel (with the DVJ in "vinyl" mode) and scratch back and forth. Now take the other and cut the sound in and out using the crossfader. There you have it: simple, unsophisticated transformer scratch technique.

Now, try it an even easier way. Rather than cut up the sound using the crossfader, cut it up using the switch that slips between "line in" and "phono" on your mixer. This way, you'll just be switching the channel on and off cold. If you're a scratch beginner, you'll find this a lot easier.

Going digital

Now you've got an option – you can rehearse the hell out of your crossfader technique, or you can let all that beautiful technology you've invested in do the job for you. Why would you do this? Why not develop "the skillz" just like the old skool boyz, rather than punch a button and get a similar effect?

Simple, because all that new technology not only lets you do things easier, it also gives you space to do new things the old skool never thought about.

Start with your EFX-1000, EFX-500 or the effects on your mixer. Somewhere, you're gonna find the transformer button. Set the transformer manually to a bpm you usually work with – in my house sets, I usually start around 128.

Make that your bpm. Now set the "transformer" effect to 1/4. That means you'll get four clean cuts every beat – or cuts in 16th notes. Apply the effect fully to the channel you're scratching on, and start scratching. It's pretty similar to what you heard using the crossfader technique above.

But where's the expressivity or complex rhythm you were able to develop using the crossfader? What if you don't want to have a wicky sound on every 16th note? Then stop moving your scratch hand!

By using the digital transformer effect, and using your scratch hand to set the rhythm (try it out and you'll understand what I mean), your other hand is still free! And that, grasshopper, is where digital scratch really starts to take off.

This transformer scratch effect is the basis of nearly all digital scratch technique. Spend your time practicing setting the rhythm with your scratch hand rather than your crossfader hand, and you'll be rockin' it in no time!

FREE-HAND MAN

There are two sides to every EFX effector – the beat effects side and the jog wheel side. You get your beat-based effects by punching up the effects and setting the bpm, either automatically through bpm sensing or manually by dialing it in (you can also get MIDI bpm using MIDI input from an instrument). Your effects include Delay, Echo, Pitch Echo, Transformer, Flanger, Filter and Phaser at a minimum, and on the

EFX-1000, you'll find considerably more effects, as well as better bpm tracking, and control from the DJM-1000 (you'll find effects details in the Effects chapter).

Choose the rhythm you want to use from the left side of the effector – 32nd notes (labeled: 1/8), 16th notes (1/4), 8th notes (1/2), 8th note triplets (3/4), quarter notes (1/1), half notes (2/1), whole notes (4/1) or tied whole notes (8/1).

If I'm dealing with the basic digital scratch effect, you would choose Transformer and then your rhythm. You can also tap a rhythm into the beat effects side, but for the moment, let's keep to 16th notes (1/4).

If you've tapped all the right buttons, you should have set up the scenario in the section above.

Now with your free hand – the one you would have

used for crossfader cutting – have a go at the jog wheel side. These effects can all be used to modify the scratch you're performing with your scratch hand. Remember that Mixmaster Mike call-in bit on *Paul's Boutique*? The one where he talked about running his turntable through a wah-wah peddle – the Tweak Scratch? This is what you can do with your freehand, rather than a complex wah-wah foot peddle setup. That and a lot more!

Try dialing up that wah effect. First, scratch with your left hand on the DVJ deck as I described above. Now use your free hand to spin the effector jog wheel left and then right (the EFX-1000 will actually continue to spin after you release it). You'll hear your scratch disappear in two different ways.

Now, this may not sound scientific, but try spinning the wheel back and forth. Really give it a go. Just go crazy. And listen to the sounds you make.

Scratch with one hand and play with the Vocoder on the jog dial. You're a scratching robot! Try the Humanizer, too.

Now try laying it all over a beat. Set the Transformer to the bpm of the track you're playing over. Get 'em synced by pressing the 1/4 button ON the beat. Now start scratching with one hand and spinning the jog wheel with the other.

Punch up another jog wheel effect and spin some more.

Again, it's none too scientific, but this is the essence of digital scratch:
- Hit the studio
- Hit the buttons
- Hit the sounds you love
- Amaze!

If you want to hear some examples of what random play can create, have a listen to a James Zabiela mix

CD. James didn't get there by learning his techniques from a book (although books can help!), he got there by playing in the studio.

And that's truly the essence of great DVJing – pushing that sense of play and pure experimentation!

Digital video scratch

If you're scratching with the DVJ – and I think you should be – it makes sense to start incorporating video scratch into your technique.

There are two routes to go here: old skool style, and digital scratch style.

Old skool video scratch involves using the T-bar or crossfader on your video mixer to cut the scratch video in and out. T-bars typically give a bit too much resistance to allow for true crossfader cutting technique for video, but some DVJs modify their V4s by installing a crossfader where the T-bar used to be.

Have a look on the net at VJforums.com to figure out how to do this. But remember, you'll not only void your warranty, you may destroy your mixer.

Another old skool method would be to take the MIDI out from the DJM-1000 and plug it into the V4 to use for crossfading. What this does is let you use the crossfader of the DJM-1000 to trigger video cross-fades on the V4. Proper old skool cutting.

The digital route is more random in terms of effect, but works similarly to the Transformer technique on EFX units. Set the BPM/CONTROL knob on the V4 to the bpm of your tune. Now hit the BPM SYNC button. Your video will fade or cut to the BPM you set up; the fade used is dependent on the Transition MEMORY effect you've chosen. Now scratch under this, just like you scratched the audio under your EFX Transformer effect.

Like it? I think it's OK, but not brilliant. The key here is as above – experiment. Mess around with several cutting styles on the V4 and see what works for you.

Spinbad and Cash Money didn't invent new scratch techniques through timidity with their machines.

A note on notes

If you're the kind of DVJ who also plays and reads music, you'll have noticed that Pioneer's method of naming notes on the Effectors is a bit – well, it's a bit counter-intuitive for normal musos. But with a bit of deciphering, you'll realize it makes perfect sense for DVJs (almost).

In standard music notation, the time signature is what tells you how to count the music. 4/4 and 3/4 time (pronounced "four-four time" and "three-four time" respectively) are the most common. 4/4 time is the norm for rock, hip-hop and house. 3/4 time is for waltzes and some good old Hillbilly love songs.

The bottom number, that 4 in both 3/4 and 4/4, tells you how many counts a whole note gets. And the top number tells you how many counts there

are to a measure.

The Pioneer system is completely different. In the Pioneer system, the top number refers to the number of beats and the bottom number refers to the number of repeats of the effect. So 1/8 means the effect will repeat itself eight times every beat. And 8/1 means the effect will repeat itself once every eight beats. Easy, right?

The discrepancy occurs with 3/4. You'd think it means the effect repeats itself four times every three beats – but what it really means is the effect repeats itself three times every beat. Go figure.

Just play with it a bit, pay no attention to the logic and it'll all be fine.

Jump in there and wail – but do it in the rehearsal studio first.

Rehearsal

I can't emphasize this enough. Once a lot of DVJs have their material produced and their beat-matching down, they do all their rehearsing in the club. Trouble is, there's no room to make mistakes or fail in the club. And if you're not making mistakes, you're not learning.

Rehearsals, and small gigs for small crowds, are the places where you can try out new techniques and see if they work. Make time for rehearsal. Then use it!

Recording your rehearsals is another route to massive improvement. What sounds right from the mixer often doesn't sound right to the listener. New DVJs often do their mix and then let the outgoing track hang around too long. This is natural, as an abrupt transition from the mixer sounds, well,

abrupt. But to the listening ear, the sudden change is far more intuitive than a long segue. These are lessons you only learn by recording your rehearsals and then listening back to them. Tunes that sound as though they work well together in the mix may sound more like a train wreck on a recording, and those you thought were disasters may be very effective. Listen and learn.

Scratch discs

While you're making all those perfect DVD music videos, be sure to whip up a few scratch discs as well. I have one disc with lots of very short tracks (as short as five seconds) that are perfect for scratching. The video motion of, say, a dancer can be effective for the back and forth of scratching, and the sound I choose is usually either white noise or a vocal. Again, experiment with different styles, but make that disc and keep it ready all night long. Drop it in when the time is right, or drop it in when you're just plain bored. But in all cases – drop it dawg! That's what a scratch DVD player is for!

DVJ ZEN MASTER

IN PARTING, I'D LIKE TO THROW AROUND A FEW IDEAS – NOT ABOUT DVJING, BUT ABOUT THINKING, AND ABOUT CREATING IN GENERAL.

I often get to speak at universities and festivals around the planet about the AV revolution, visual art, the internet, my work and just about anything else they'll let me get away with. One of my harangues, particularly when I'm speaking to uni students, concerns technique – not its importance, but rather the lack of importance of technique in the 21st century.

Unless you're a dancer, a violinist or a footballer, in the face of new technologies, technique has become a secondary consideration. Technique is only a starting point – a thing to master, in many cases, but only in aid of freeing you from the restrictions of technique. What matters most is what and how you think about what you do (actually, this applies to dancers, violinists and footballers as well).

Dancers and violinists are straightforward examples. Both artforms arrive with enormous technical baggage. Instrumentalists are schooled in the canon of music history and traditional technique. Dancers must learn technique in order to control their bodies. But great artists in both traditions spend lifetimes learning to unlearn technique, in order to arrive at some competent level of expression.

The same applies to contemporary music. If Miles Davis or Ornette Coleman had spent their lifetimes playing the way they were supposed to play, they'd never have invented either Jazz Fusion or Harmolodics.

Whether discussing the work of Sigur Ros or Glenn Gould (the 20th century's greatest interpreter of Bach), the central point of the conversation will turn toward a "turning away from traditional technique". What is at stake is the moving beyond of what was considered appropriate for either pop or classical performance and, particularly with Gould, recording. To achieve a break like this requires an ability to think about one's own medium, beyond the traditional techniques.

In some schools we would call this theory, and in others it has been called contextual study. Let's just call it thinking. You must learn to think about DVJing. Without critical skills – an ability to conceptualize your practice – you are bound to repeat what is done by other DVJs. It is nearly impossible to discover your position within a field if you don't know where the borders lie.

That said, I'd like to throw around some ideas for you to mull over. I have more, but I'm running out of space. And, if after a bit of thinking, you'd like to come back to me with some of your own, please do so, by all means. Just email dvj@kriel.tv – I'll get back to you.

The linear mistake of the 20th century

Prior to the invention of the phonograph in the late 19th century, music was far from a beginning-to-end recorded experience. Many households held at least one musician, and music was an experience of interaction, either with other musicians or with written sheet music – the 19th century's primary form of music publication (there were also, of course, piano rolls and music boxes). Sheet music left enormous room for interpretation by the individual. And even pianolas, of the non-automatic variety, required a large degree of interaction from the player, who controlled dynamics and tempo (pianolists were, in many ways, the first DJs). In other words, the audience for most recorded (printed) music was in fact the artist, who in turn performed before and interacted with an audience.

Something went wobbly with Edison. Music was inscribed in a soft material like wax, but it was also inscribed in a form that could not be read by anything other than a machine. Rather than a set of encoded instructions to a performer – notes on a page – music became a squiggle of vibrations to be read by a needle attached to a funnel. And the only option was to play it, unfortunately, from front to back. That last bit is crucial.

In the 20th century, much of what had previously passed as work that interacted with an audience – storytelling, music, etc. – became rigidly locked in a front-to-back, non-interactive fashion. The code of the work could only be read by machines, and audiences were asked to sit still and listen until the record or program was over. Granted, you could clap or dance, laugh or heckle, but it would result in not one iota of change – the humorless plodding machine would continue on its path to the end of the work. After records, we got film, radio and television, but all shared this one feature – all were transcribed and read by machines, and therefore forced the public to politely fold their hands in their laps, shut their traps and act like an audience.

Fortunately, computers and non-linear media have changed all of that. Although audiences today, for the large part, still operate within the constraints of an author's vision, we are much more free to get up and move around. No longer are we forced to watch James Bond on the screen – we're now given the opportunity to pretend we're James Bond and pursue a path with at least a few choices on the Playstation. Indeed, many artworks today even ask us to make the work ourselves.

The DVJs, along with software like *Ableton Live* and *Grid Pro*, invite us to disrupt the linearity of the media we've been fed for a century. We're asked to cut things up, layer and shape them, and create our own works out of the works that surround us.

Although the 21st century is no media utopia, and indeed the good folks of the last century got along just fine without an Xbox, thank you, things have changed substantially. The artist formerly known as the audience has taken up a degree of control, and with it responsibility, for their creative lives and their cultural surroundings. But more important, we have been propelled forward into an era that exists after linearity. Front to back is far less important than left to right and everywhere in between. The 21st century's greatest storytellers will be the creators of interactive experiences (like the club) and games. So, my question is this – how will you respond to this as a DVJ? Think it over.

Picking future hits and trends

A&R men from most major record labels have embarked on a peculiarly 21st century practice to help them choose not only which tunes they think will be hits, but which acts to sign. They use a piece of software that analyzes the melodic, lyric and structural patterns of potential releases and returns a result that predicts the likelihood of a hit.

I'm not kidding. The process is called HSS – for Hit Song Science – and the software was developed by a Spanish company, Polyphonic HMI. Some record executives treat HSS results as gospel, but some are less sanguine. In an article in *The Guardian* by Jo Tatchell, Mike Smith, A&R director at EMI, said "A good A&R has a very accurate instinct for what the market needs." In other words, a good A&R man is using HSS to justify to his boss what he already believed to be the case.

In the same article, Tom Findlay of Groove Armada put it this way, "... (W)hile there are rules of construction in play – verse, bridge, chorus and so on – the aim as a musician is to make the musical statement you want to make. The end game is not to get the mathematics right."

I think he's right, but I also think there is more to it than that. Personal expression is very important for an artist, and an understanding and sympathy for that expression is one of the key talents of good A&R. But how do you pick a hit? How do you add a bit of science to the instinct that tells you, "When I drop this track, the place is going to go off."

HSS works by evaluating the patterns of a tune and comparing them to the patterns of hits from the past. Although I hope there's more to it, that is roughly HSS's methodology.

So, we have personal artistic expression and similarity to that which has worked in the past. What else is there? Frank Zappa put it succinctly – timbre. He said pop music is all about timbre – the quality of the sound. Detecting timbre is the way you can tell the difference between an electric guitar and a banjo, just by listening. Melodies, structure, chord progressions – none of these have changed significantly over the past few decades. But timbre is in constant flux. Hemlines may rise and fall, but they're still hemlines. What makes a dress fashionable is how it goes up and down – the cut, the material, the quality of the look. The same goes for fashion in music – it's the timbre

that is the principal determining factor.

In the early noughties, a guitar solo was a guarantee of disaster in the middle of a pop tune. Was this because we were tired of melody or instrumental breaks? No, we were overloaded with wailing single-note, high string electric doodling from the blow-dried guitarists of 80s and 90s rock. So we stopped listening and did without guitar solos for 15 years. Guess what? Here, in the mid-noughties, they're starting to sound kinda fresh.

The same goes for House beats. All the rage around 2001, they're a bit stale at the moment. Not that we're not interested in four-on-the-floor, or that 128 is an unpleasant tempo, we'd just like them to sound that little bit sharper, a little more live. We'd like the timbre to shift.

Musical fashion is determined by timbre. And musical success – creating a "fresh" sound – is contingent on creating a sound that is a little like what came before it, but just different enough to be exciting. And this, my friends, is beyond HSS.

The same applies to visuals. The 90s brought a wave of "noisy" video and stills – lots of film grain, cut and tear, spilled ink and smudged colors. The early noughties, in response, embraced clean lines, solid blockish color and rigid animation – a la vector graphics. And a bit of nostalgic pixel animation (see eToy). The mid-noughties, in turn, have embraced the flourish – a complex, fluid type of illustration, with intricate but repetitive patterns, intertwining lines like vines on a tree – nearly art nouveau, but with sharp if somewhat delicate edges. And colors of significantly less contrast. Not quite noise, but not quite Flashesque computer animation. Just a little different than what came before, but still similar.

Now two questions: first, what do you think will come next; and second, how will you respond as a DVJ?

In praise of doing it wrong

My computer helps me do everything the right way. I can correct my vocals, sort my harmonies, quantize my beats, clean up my photos, check my spelling and get rid of my red eyes (as well as attempt to predict a hit song). But my computer can't – crashes aside – do things wrong. You can, and that frees you from the machine. If you find yourself stuck in the creative process, try to do it wrong. Often our greatest expression comes via confrontation against our limitations – whatever they might be – but our greatest inspiration often arrives by embracing those same limitations, or by the influence of others who do. I find inspiration in Will Oldham, Harry Partch, Charles Ives, Stina Nordenstam and the strange melodies of an unobserved five-year-old singing to themselves – even though, most of the time, I play beats.

As a DVJ, where do you find yours?